Llama Keeping

Raising Llamas – Step by Step Guide Book… farming, care, diet, health and breeding

By Harry Fields

Copyright and Trademarks

Disclaimer and Legal Notice

Foreword

In the mid-1980s, I had my "mid-life crisis" several years early. After struggling through the politics of a graduate school program, I decided that there was too much backstabbing in academia for me. To the great shock of my friends and family, I bought a few acres in Colorado and moved to a tiny cabin on the land.

I confess there was a plan to write a "great American novel." Isn't there always? What I wound up doing was reading for hours, taking up fly fishing, learning to chop wood, and sitting on the porch – a lot. It wasn't great for my bank account, but it was fantastic for my very frayed nerves.

Sometime in the first month, when I was sitting out on the porch, I started watching the "sheep" in my neighbor's pasture. I remember thinking, "Those are some really long necked sheep!" Finally, my curiosity got the better of me, and I strolled over to our shared fence line. The "sheep" were llamas, about half a dozen of them.

The animals didn't bolt at my approach. In fact, they stood placidly chewing their cud, regarding me with frank interest. As I leaned on the fence, my neighbor pulled up in his beat-up old truck and asked me how I liked his "girls."

I admitted that I had never seen a llama before and was promptly invited to supper and a tour of the farm. The promise of food other than my own cooking was as powerful motivator, so I accepted with both interest and gratitude.

That evening was not only the beginning of a lovely friendship with my neighbor and his family, but also an introduction to llama care. I wasn't making much progress with my great American novel, and my neighbor, probably out of a sense of charity, asked if I'd like to help out around the place a little bit for pay.

I grew up in a small town where there were lots of farmers and ranchers. I wasn't afraid of livestock, but I didn't have any hands on experience with them. My "country" knowledge extended to polite dictates like, "If you open a gate, close it."

Over the next couple of years, I walked or drove over to my neighbor's several times a week. He taught me many things, including the rudiments of llama care, which is why I say – more than once in this book – find a mentor.

There's more to caring for animals of any kind than just following a set of instructions. You have to cultivate a sense of the animals' personality and view of the world.

What you might do with a cow would never work with a horse, and you can't treat llamas like what I thought they were, long-necked sheep.

Please understand that I'm not attempting to write the definitive book on llama care. I am interested in providing you with a comprehensive introduction so you will be in a position to ask better questions and to learn about these fascinating creatures in a hands-on setting.

I wish I had known a lot more about llamas from the start, because I assure you, I was the butt of many good-natured grins and jokes. My neighbor's wife called me "Professor Old McDonald," to which I always answered, "e-i-e-i-o." It was all great fun, but I really was horribly ignorant.

Ignorance is, however, completely correctable. I hope when you are done with this book you'll know enough about llamas to have a sense of where you want your interest to lead. Maybe you've come to these pages with a minimal curiosity that I will satisfy. But maybe you really do want to bring llamas into your life and you'll be prepared to take the next step.

Absolutely no living creature should be purchased or adopted on a whim. All animals deserve a high standard of care. Llamas are easy as livestock goes, but they still have real needs and specific requirements to thrive.

Eventually I did take up writing as my full-time vocation, and ultimately I sold my little patch of Colorado country and returned to the city. I have many fond memories of my time there, and those memories include the llamas next door. I'm glad caring for them has been a part of my life.

If you decide to weave that thread into your own personal tapestry, I can promise you, you will be enriched by the experience. Llamas are gentle, curious, intelligent, and very, very interesting.

What you'll discover inside…

Would you like a step by step guide on raising llamas? Read through this book and you'll discover everything you need to know from…

- The best type of llama shelter (and why)
- Must know feeding advice that most books don't include
- What important things you need to know about worming
- What fencing is recommended and what to avoid
- How to recognize and more importantly how to prevent some common health problems

And lots lots more…

Selecting a healthy llama, associated costs, housing, husbandry, health and breeding, plus useful frequently asked questions. Each section is covered in detail. You'll also find a breeder's directory detailing sites across America and the UK plus a list of relevant websites.

What Readers Have Said

"Highly recommended book. This is probably the best book I've read on Llamas and I've read a few. It's well written, easy to follow and covers the information you need to know about looking after these amazing animals" Karen Slade

"Anyone who is looking after (or wants to buy) llamas should read this book. There is so much great advice. I found the health and feeding sections particularly useful" Agnes Naylor

Table of Contents

Table of Contents

Chapter 1 - Introduction to Llamas

The Llama (*Lama glama*) is a camelid indigenous to South America. As a member of the *Camelidae* family, they are related to alpacas, vicunas, guanacos, dromedaries, and Bactrian camels.

In physical form, camelids are large mammals. Many, including the llama are big enough to serve as pack animals. Camelids are herbivores and quasi ruminants with three-chambered stomachs.

Their three-phase digestion begins with consumed plant matter passing to the rumen, where a process of microbial fermentation occurs. Next, the partially digested food moves to the reticulum. There the solid and liquid materials are separated. The solids are clumped together to form a bolus or "cud," which is then regurgitated and chewed again before going to the abomasum, which is the true stomach.

Camelids are also even-toed ungulates, using the tips of their toes or hooves to sustain the bulk of their body weight when they are in motion. Other large mammals with the same foot structure include (but are not limited to) pigs, cattle, deer, sheep, and goats.

The llama, alpaca, vicuna, and guanaco are New World animals, while dromedaries are Arabian and Indian camels. They have the traditional "camel" appearance and a single hump. The Bactrian camel, indigenous to Central Asia, is a shaggy, short, domesticated pack animal with two humps.

Background and History

Ancestors of the modern camelids lived on the plains of North America 40 million years ago, and were likely forced into South America during the Ice Age. The major camelids seen in the Americas today are the alpaca, vicuna, guanaco, and the llama.

Alpaca

Alpacas, which are often confused with llamas, are grown primarily for their fiber, which is classed in as many as 52 colors in its natural state. Adult alpacas measure 32-39 inches / 81-99 cm at the withers or shoulder, and about 5 feet / 1.5 meters from the ground at the head. Adults weigh 106-185 lbs. / 48-84 kg.

Vicuna

Vicunas have long, wooly coats. The animals are brown on the back and white on the throat and chest. They live exclusively in South America and have not been domesticated. At the shoulder they stand about 3 feet (75-85 cm) tall.

Vicunas were declared endangered in 1974 when there were only about 6,000 left in their native Peru. Due to the hard work of conservationists, their numbers have now climbed to 350,000. The greatest ongoing threat to the national animal of Peru is now habitat loss.

Guanaco

Guanacos are similar to the vicuna and are also wild animals. They are slightly larger than the vicuna, with gray faces and bodies that range from dark cinnamon to light brown. The underside is white. With fewer than 600,000 guanacos left in the mountainous ranges of South America, they are regarded as vulnerable from a conservation status.

The species is still considered wild, but approximately 300 guanacos are now placed in zoos in the United States with an additional 200 registered in herds kept by private entities. The guanaco is classified as the parent species of the domesticated lama.

Llamas

Llamas are graceful and elegant animals with beautiful wool coats that range from white to black with all shades in between including grays, brown, red, and roan. You will see a wide variety of markings and patterns as well as solid-colored animals.

A fully-grown llama stands about 5.5-6 ft. / 1.7-1.8 meters at the top of the head and weighs in at 280-450 lbs / 130-200 kg by age four. There are no great differences in the genders although males are slightly larger.

There are four basic coat types:

- Wooly Llamas have a strong coverage of wool over their entire bodies with greater concentrations at the head, neck, ears, and legs. They produce fine, crimped fiber on par with alpaca wool in quality and exhibiting very few guard hairs.

- Medium Llamas have longer fibers on the neck and body, but shorter coverage on the head, ears, and legs. They have rough protruding guard hairs and are thus said to be double coated.

- Classic Llamas have less wool on the head, neck, and legs with slightly longer hair on the back forming a saddle down the sides. The guard hairs on the neck appear to be almost a mane.

- Suri Llamas are extremely rare and difficult to breed, but their wool, like that of the Suri Alpaca, hangs in rope-like tendrils and is less fine than that of the Wooly Llama.

Llamas eat by grinding their upper and lower molars back and forth. They have no upper teeth in the front, but use their upper lip to help them grasp food with their lower incisors.

Adult males develop large and very sharp fighting teeth that must be ground back to the gum line to prevent serious injuries during territorial scuffles and general roughhousing.

Their two-toed feet are well adapted to make them sure-footed on all types of terrain from sand to snow. The foot has a leathery pad at the bottom and there is a curved nail extending from each toe.

Llamas are intelligent herd animals with a lifespan of 20-30 years. They can serve their human masters in many ways. Llamas produce a very soft, lanolin-free fiber and as a beast of burden, they can carry 25%-30% of their own body weight for a distance of as much as 5-8 miles / 8-13 km.

Because they are territorial, and brave enough to be aggressive in defense of their "herd," Llamas make excellent guard animals for other types of livestock.

(Please note that there is some controversy over this accepted "fact," which I will discuss later in this book.)

Miniature Llamas

In the United States over the last 20 years breeders have paired small adult llamas to create a miniature version of this popular animal. Currently the gene pool for miniatures goes back 6-7 generations and still does not produce consistent or reliable results.

However, this fact did not stop the International Lama Registry from recognizing miniature llamas as a separate breed in 2005. Regardless, many skeptics in the industry refuse to accept the animals as anything other than smaller standard llamas.

According to The American Miniature Llama Association for an animal to qualify as "miniature" it should stand no more than 38 inches / 96.5 cm at the shoulder by age 3. This is 75% the size of a standard llama.

Currently there are only about 1,000 miniature llamas, most in the United States. The notion of a smaller llama is appealing for a number of reasons, including easier management. Miniature llamas eat less than full grown llamas and as many as six can forage well on a single acre of land.

You can feed a miniature llama for roughly the same amount of money needed to keep a large dog, and the llama will mow your grass for you! The smaller animals are easy for children to handle, and are known for their gentle temperament and high intelligence.

Miniature llamas can be used successfully as therapy animals in hospitals and nursing homes, and are suitable for people with disabilities. Some people even bring their mini llamas into their homes!

Although they cannot carry as much as their full-sized cousins, miniature llamas can still work as pack animals and will happily help you carry your gear on a hike for a picnic in the mountains or an excursion to a remote corner of the beach.

In every other way except size, miniature llamas are identical to standard llamas. They are harder to find, but if you have limited space and are primarily looking for a pet, one of these little beauties could be just right for your circumstances.

Differences in Llamas and Alpacas

The easiest way to tell an alpaca and a llama apart is to look at the ears. Llamas are physically larger than alpacas. They have curved ears that look like bananas, while alpacas have straight ears.

The coat of a llama is interspersed with coarser guard hairs that must be removed during processing, while alpaca fiber is very fine and soft. It is a myth, however, that llamas do not produce wool suitable for use in fine textiles.

The coat of the Wooly Llama is almost as soft as that of the alpaca and shares its hypoallergenic properties, minimal weight, exceptional warmth, and lack of lanolin. The prickle factor of both fibers is very low, so neither cause the itching reaction so prevalent in sheep's wool.

Behaviorally, llamas are much more aggressive and are often used as guard animals for grazing herds of sheep and even alpacas themselves. Not all llamas are suitable as guard animals, but females working in pairs can effectively protect livestock from small predators.

Potential Commercial Uses for Llamas

Individuals interested in exploring the commercial uses for Llamas will find that they are highly adaptable animals with an inherent ability to "multi task." It is quite possible to own several llamas and to use them in different ways according to the characteristics and "talents" of each animal.

Pack Animals

Over the past 20 years, the use of Llamas for commercial and recreational packing has been on the rise. They are quiet animals that literally walk and live gently on the land.

Llamas do not pull up the plants on which they graze, only snipping off the tops, and their delicate feet don't harshly effect the trails on which they walk like the hooves of larger pack animal such as a mules or horses.

For this very reason, horses and mules are banned on some protected federal lands within the United States national parks system, but llamas are allowed.

The ruggedness of the terrain is not an obstacle for sure-footed llamas, but they do not do well in harsh temperature extremes. They are, after all, indigenous to the high mountainous regions of South America.

As companions for hikers, backcountry fishermen, and even workers like surveyors, llamas are useful and congenial companions and beasts of burden.

Fiber

Although llamas do not enjoy the rarified status in the world of fiber production held by their relatives the alpaca and the vicuna, interest in llama wool has been on the rise since the 1990s.

Llama fiber is excellent for spinning and felting, and is particularly useful when combined with sheep's wool for added sheen and luster. Llama fiber is warm, lightweight, and does not shrink.

One of the greatest advantages, however, is that the material, which contains no lanolin, is hypoallergenic. It can be used to stuff pillows, and to produce garments that do not have the prickle or "itch" factor so often seen with sheep's wool.

Fertilizer

People who have no experience with camelids are surprised to discover that both llamas and alpacas are amazingly clean animals. Both use communal dung piles in their pastures, which makes clean-up and maintenance extremely simple and efficient.

For this reason, anyone keeping llamas has the option of harvesting the odorless manure pellets for use as a fertilizer. The material is high in nitrogen content, but it will not burn plants, and does not necessarily require composing.

Fertilizer from llama manure can yield on average around $10 / £5.93 per 2 lbs. / 0.9 kg. The quality of the dung is approximately that of bat guano, which has a 4-3-2 nitrogen-phosphorous-potassium ratio.

Livestock Guard Animals

A llama's highly territorial nature makes some of these creatures well suited to serve as guardians for other types of livestock. It's impossible to tell, however, if a llama might be a suitable guardian until the individual has reached 18-24 months of age.

There is a growing industry for identifying and training guard llamas and instructing livestock owners on their use and introduction with other species.

Do not think that you can just turn any llama loose in the field and expect it to guard your livestock. You could go out the next day and find your sheep *and* the llama dead from a predator attack.

Some factors to consider in looking for a guard llama include:

- Age 18 months at minimum. Younger animals lack the size and strength to ward off predators, nor are they mature enough to assume a responsible guardian role.

- Health and conformation. Obviously the llama should be in excellent health and have no physical defects that would prevent it from performing athletically while "on the job" or foraging well in the pasture.

- Gender and intact status. Traditionally gelded males have been sold as guard animals so long as the gelding was performed after 18 months of age and before 3 years. Conventional wisdom now holds that both intact males and gelded llamas present too high a risk of indiscriminate mating behavior. For this reason, pairs of females are now favored to work as guards.

Generally, successful guard llamas in addition to being physically sound have a natural curiosity that leads them to be unusually aware of what is going on in their surroundings.

The more hypervigilant and engaged a llama is, the better "watch dog" it will be. You do not want a bored and nonchalant llama in charge of your sheep.

Llamas that do not like to leave their sources of food and water, prefer to be near the barn, panic easily, and are generally timid are also not good candidates for guarding jobs.

Don't pick any llama that displays aggression toward humans. The animals should be halter and lead trained and load easily into a trailer for transport. If the llama is to work in conjunction with a guard dog, the animals should be introduced to one another in a controlled environment. Not all llamas are tolerant of such partnerships, however.

As Pets

It is perfectly plausible to keep a pet llama (or preferably a pair of llamas). They are actually less expensive to maintain than a family dog, consuming only about a bale of hay per week and happily grazing in pastures during the warmer months.

They are calm and tend to be very good natured, and are easily handled by children (which also makes them popular show animals for youngsters involved in programs like 4H and the Future Farmers of America.)

General Observations on Keeping Llamas

Due to their low maintenance profile, hardy and disease-resistant constitutions, non-challenging behavior with fences, and overall manageability, llamas are highly attractive animals to be kept for almost any purpose.

It's a mistake, however, to think of owning llamas as some kind of get-rich-quick business. Very few people – only about 5% of the industry -- keep these animals as a full-time business.

Profitable llama operations involve substantial financial investment. In addition to acquiring quality livestock and providing them with a high standard of care, the business must be marketed and promoted according to its focus. There is, for instance, no point in offering outstanding stud services if no one knows about your operation.

There are clearly many models for llama ownership, and each one can become the foundation for a business, which is part of what I hope to explore in the remainder of this book.

In practical application, however, most people who have llamas engage in some combination of these activities. As you will soon learn, llamas are very versatile and can help "pay their own room and board" in a variety of ways. First, however, let's talk about best practices for selecting and buying llamas.

Chapter 2 - Buying Your First Llama

Obviously a llama is not the kind of animal you purchase on a whim on a Saturday afternoon – or it certainly shouldn't be! You have a great deal to learn in advance of your first purchase. This education should involve reading (which you are clearly doing), joining llama organizations in your area, and spending time on llama farms or ranches.

Visiting Farms and Ranches

Visiting farms or ranches where llamas are kept and getting to know people who work with these animals routinely is an absolute prerequisite for getting into any form of llama-based business. There's only so much you can learn from reading books or even watching videos. You need friends and mentors in the business!

Look for llama shows in your area. The animals are often shown at state fairs or as part of larger livestock shows. Approach breeders and collect business cards. Make sure the breeder has time to talk with you and explain that you are interested in learning more about the animals and potentially becoming an owner.

I think you will find that people who show llamas are more than happy to talk llamas day and night! Ask to visit their farm or ranch so you can get some hands on experience (or at least "eyes on" time) with the necessary equipment, shelters, and feed.

Talking with a llama owner will give you a chance to ask about daily care, vet bills, insurance, tax breaks from owning livestock (these are available in the United States, but tax codes vary widely by country), and anything else that comes to mind.

Take notes and ask permission to take photos. Reading over your impressions and looking at the images later will lead to more in depth questions and will give you a frame of reference for thinking about erecting shelters and fences on your own land.

Try to cultivate a relationship with your new llama-owning friends so you have someone to turn to for advice, especially after you purchase your first animals. Once your llamas are "home," all kinds of things will come up that you never anticipated.

On a whole, I think you'll discover that llama owners are a close-knit and convivial bunch of people. Learning the industry from the inside from people who have already been there and made the "newbie" mistakes is absolutely your best bet.

This kind of mentoring arrangement gets around any notion of absolutes and helps you learn to respond to the specific needs of your animals and of your unique operation.

All of the information that I will provide for you in this book should be considered a basic primer. Use this foundation to expand your understanding of the animals and their husbandry and to spur your further research.

I cannot stress strongly enough that you need to engage in this course of self-education BEFORE you purchase any llamas, and that all of the equipment and shelters for their proper husbandry be in place BEFORE the animals arrive on your land.

Do I Have to Own Land?

For as obvious and even silly as that question might seem on the face of things, I hear it all the time. "Do I have to own land to raise llamas?" The answer is, "Yes."

Llamas are not backyard animals, and if you try to keep them that way, you will find out about local zoning laws and livestock ordinances very quickly – most probably with a fine attached.

If you live in a community with a homeowners association, I can promise you that raising livestock in the backyard is against their covenants!

One model of llama ownership that was made popular in the 1970s was an arrangement whereby individuals or groups of people invested on a share basis in llamas that were kept at a central location.

This was, however, a short-lived fad. Certainly an arrangement of this type within a family or group of friends is not impossible, especially if one person in the circle does own a "place in the country."

But for the basis of this examination of llama husbandry, we are going to assume that you have land in the country and will be erecting shelters for animals that will graze in pastures.

Starting Small

Many books on llama husbandry will counsel newcomers to the industry on the virtue of starting small. I do agree with that, but if possible, I think it is best to have a pair of llamas rather than a single animal. They are, after all, very social creatures and have a real need to interact with others of their own kind (or even other livestock.)

Llamas that are well socialized are much easier to handle. They tend to be friendlier, and to even have a nurturing and almost whimsical attitude toward their keepers and the world at large. Solitary llamas can become lonely and frustrated, taking out their feelings in aggressive behavior.

If you get a single animal, starting with a gelded (castrated) male is a good way to get experience with the species. Most single llamas, regardless of gender or intact status can be kept with other livestock, but do not house intact males with female sheep and goats.

(Please read Chapter 9 – Llamas as Livestock Guardians for a more complete understanding of the relationship between intact status and behavior.)

Your "starter" llama will stay with you no matter how your business develops or what your goals become. That first llama has a way of remaining the favorite. After all, he or she took on the responsibility of breaking *you* in to the business!

Choosing Animals

Depending on the envisioned purpose for keeping the animal, you may apply different criteria to choosing a llama for purchase.

For instance, if the animal will primarily be a pet, but will be shown by your son or daughter as a 4H project, you want a llama with a superior disposition and personality. If you plan to use the llama as a pack animal, you want a body conformation that is more heavily boned and taller.

Regardless, however, there are some general points to consider in picking any llama.

- Does the animal appear to be healthy and are their records of any treatments, including vaccinations that it has received?

- Is the llama's back straight and are the neck and tail set at good angles from the body?

- Are the legs straight and are the ankle joints vertical rather than dropping toward the ground?

- Does the animal seem calm and does it react well to handling?

- Has the llama received any training? If so, what kind?

- Is the animal used to wearing and being handled with a halter?

- Does the coat appear to be healthy?

First time buyers should always work with someone who is willing to offer help, generally in the form of advice and information, after the purchase if it is needed.

Bringing your first llamas home is a little bit like bringing a baby home. You've read all the books. You think you know it all – and then the unexpected happens and you panic and need a more experienced "parent" to help you out. That's perfectly normal.

Since the prime consideration is always the welfare of the animals, most llama owners are more than willing to help out newbies to the industry. As you become more experienced, adopt an attitude of "pay it forward" and help the next new person down the line.

Evaluating Health

When you are considering buying a llama from a farm owner, make an appointment to see the available animal(s) and arrive on time. Take in all the information you are receiving, both in conversation with the owner and what you can see around you.

Observe all of the livestock on the property and the condition of the facilities in which they are being kept. The animals shouldn't be skinny, but neither should they be overweight.

Ask the seller what vaccination and worming practices he follows. What specific vaccines and worming agents does he use? Why those products? How often are the llamas specifically vaccinated and wormed?

Ask to see the registration papers for the animal in which you are interested, as well as its health, vaccination, and worming records. Production records are also important if you are considering buying an intact animal.

If you are buying a female find out what kind of mother she is and if she has had any problems giving birth. How have her offspring performed? Are they fiber animals? Pack animals? Is she capable of producing cria that will grow into animals that will fit your breeding and business goals?

Especially when you are just getting started, ask if you can come back to the seller with questions after the deal is concluded or if problems arise. If there are guarantees attached to the sale, make sure you receive those in writing.

As in any other kind of business deal, learn to trust your intuition. If you have an uneasy feeling, or something just doesn't seem right, thank the person for their time and move on.

Transportation Considerations

Unless you are extremely well equipped and have experience transporting llamas, it's better to hire a livestock transporter to move your animals over great distances. This is especially true if you've just purchased llamas in another state and need to get them to your land safely.

Obviously as your experience with the animals becomes more extensive, and depending on the needs of your business, you may well acquire the right kind of trailers and become adept at managing these trips yourself. As a new owner, however, this may all be completely overwhelming.

Get estimates from 3-4 haulers since their rates will vary widely, and make sure you ask for references and actually check them out. I would not, however, be motivated by price alone. Make sure that the hauler with which you contract is experienced and capable of protecting your animals in transit.

Llamas are easily stressed. You want to choose a hauler who knows the animals well and understands what is required to keep them safe and well on a long trip.

Don't expect to get a direct delivery unless you're willing to pay for it. Typically livestock haulers book deliveries all along their planned route and drop animals off along the way. Your llama could be in transport for days.

Both the seller and the hauler should be able to advise you on the required tests and health certifications for interstate or international shipping. Make sure the hauler has all the necessary documents in hand and that you have copies.

There are a number of things you want to find out in advance. Be suspicious of any haulter that isn't willing to address these concerns or that is evasive in any way. These questions include, but are not limited to:

- Why type of trailer and truck do you use?
- What do you do to safeguard the animals if you break down?
- How do you deal with sick and injured animals?
- If you are carrying other types of livestock, will my llamas be kept separately behind partitions?
- Will the llamas come into contact with any species en route including cattle, sheep, and goats from which they can contract diseases or parasites?
- Can I send specific feed on the trip or do the animals have to eat what you provide?
- Will you be in contact with me if something goes wrong?
- Will you let me know if the trip is delayed and behind schedule?
- Will I be provided with a cell phone number so I can contact you en route?

Llamas that are well seasoned to travel can be carried in a van or an SUV, but in most cases after purchasing an animal, it will be delivered to you in a traditional horse or livestock trailer.

As soon as a vehicle or conveyance in which they are placed starts to move, llamas prefer to kush, which is their resting position with legs tucked under their body. Make sure there is good ventilation, or if possible, use an air-conditioned trailer.

Animals can die quickly from carbon monoxide poisoning. It is crucial that no exhaust from the engine get into the area where the animals are resting. The trailer should also be outfitted with nonslip trailer mats.

Buying Advice Summary

Just to summarize the major point on buying llamas, remember the following.

Buy animals that are appropriate for your goals.

If you are interested in owning pack animals and offering their services, then buy llamas with the correct physical conformation that have been trained for the work they will perform.

Don't expect a guard llama to placidly walk up the trail obligingly carry a load. He's been raised and trained to guard and that's what he's going to want to do.

Don't fall for sales tactics.

If you approach a llama breeder and say, "I want to buy llamas to produce fiber," and his answer is, "I have some *great* pack animals to show you," don't fall for it! Set your business goal in the planning stage and pursue that goal with focus.

Be realistic about the value of your purchase.

Obviously budgets are important, but if you are offered a llama that is well trained and has excellent conformation, be willing to pay a little more.

You can always ask for a pre-purchase veterinarian examination to confirm the animal's quality. The better a llama's conformation, the less likely it will be to develop leg problems and arthritis among other issues.

Seek expert advice if you need it.

If you don't know what you're doing or if you're simply unsure about your first purchases, find someone or hire someone with the knowledge you lack to locate good quality llamas and to help you negotiate the terms of the transaction.

Start small, with a pair of adult llamas.

This is a point that cannot be repeated enough. Start small. Give yourself at least a year to learn and to find out if you like working with llamas well enough to really make them your business.

This is especially true if your operation will include breeding. You have a lot to learn about genetics as well as husbandry to produce good quality livestock! Always start with adult animals. You're not ready to take on the care of a young, developing cria.

Be a discriminating shopper.

Make sure that you are dealing with reputable breeders selling registered animals with verifiable bloodlines. If you are being told that the animal you are considering has been trained in some way, ask for proof of this fact.

Don't buy llamas that you can't catch or that haven't been well trained. You won't have the necessary experience to work with them and control them, and they'll be a major headache -- not to mention the stress the animals will suffer.

Chapter 3 – Understanding Llama Husbandry

Anyone familiar with keeping other forms of domestic livestock will have an easy time learning good husbandry practices for llamas. In fact, you may readily understand that llamas are easier to care for and more economical than many other types of livestock.

The following information is intended to give you a good start in caring for your animals, but truthfully the best way to become skillful with llamas is to work with them under the mentorship of someone who knows the animals well. And yes, I know I am repeating myself, but it's a point I am over-stressing on purpose.

If at all possible, spend some time on a working llama farm before you try to take care of your own animals on your own. Literally get your hands dirty. The welfare of living creatures depends on your understanding of their needs. You must be well prepared *in advance*.

General "Best Practices"

As a matter of general precaution, if you are uncertain about a llama's health condition, isolate the individual in sight of, but separate from the rest of the herd until you can obtain sound veterinary advice. Typically if serious illness or a decline in health has not manifested in two weeks, it's safe to reintroduce the llama to the herd.

Always have newcomers to your herd evaluated by a veterinarian and make sure a fecal sample is taken to determine if the animal needs to be wormed. Watch all your llamas to make sure they are eating, ruminating (chewing their cud), and passing feces in the form of firm pellets.

In their indigenous range in the high Andes, llamas live in arid conditions, but they are so adaptable as grazers and browsers, they can be kept in a variety of temperate environments.

Highly efficient digestion and low protein requirements allow llamas to eat only 2-4% of their body weight in dry matter daily. If no pasture is available, a single 100 lb. / 45.35 kg bale of hay will feed an adult llama for 10 days to 3 weeks. When pasture is available, 3-5 llamas can graze per acre.

Grain should only be fed to pack animals and nursing females so long as good hay is available. Provide sheep mineral and salt blocks as a free choice option, but switch to granulated minerals as necessary since llamas often find it difficult to lick blocks.

The restriction on high-protein grain mixes used so commonly with other forms of livestock speaks to the llama's tendency to bloat. Don't let them get into an unsecured grain bin!

Create a regular schedule to clean dung piles and to administer preventative medicines like vaccines and deworming agents. (See the chapter on health for a more complete discussion of these healthcare procedures.)

The Importance of Good Quality Hay

Quality hay is critical to keep your llama healthy. You'll need to educate yourself about hay in greater detail, learning about factors like protein content and digestible nutrients. Often this is a matter of judgment in relation to what is available locally.

If, for instance, you have access to baled coastal Bermuda that has been stored well, it has good protein content and more digestible nutrients than first cut alfalfa that has been rained on.

Fescue hay that has been grown in the United States or Canada may be infected with entophytic fungus. The alkaloids it produces are toxic to many animals, especially during pregnancy. Although this toxicity to llamas has not been confirmed, it's best to avoid fescue in favor of safer options.

Good hay doesn't smell sour or musty. It should smell fresh and clean, and be free of mold, dust, and weeds. There should be no sign of saplings or flowers, or any other type of debris that got into the material during baling.

Always look for any signs of insect activity and reject the bales immediately if you find bugs. Many insects carry serious diseases. Better safe than sorry.

Reject any bales that are unusually heavy or that feel warm to the touch. This is an indication that the hay was not completely dry before it was baled.

The heat is generated by the dampness, which will also trigger mold, and in some instances even spontaneous combustion in storage!

When a bale of hay is opened, it should be green inside. If it is brown or yellow, the hay was rained on in the field or bleached in the sun before it was baled. Both will lower the nutritional value of the hay.

There is no need to be concerned about discoloration on the outside of the bale, which is normal, especially if the hay has been stacked outside in the sun.

Alfalfa hay is often neon green because it has been treated against mold with the preservative propionic acid, a naturally occurring organic fungicide. It is perfectly safe for use with livestock.

If you can't find superior quality hay like Bermuda, bagged grass or alfalfa works well.

Evaluating Hay

As a good option to find high quality, tested hay, visit your local agricultural office and ask for a list of hay suppliers in the area. Don't go by word of mouth. It's too easy for bad hay to be passed off as acceptable and sold for premium prices, especially to unsuspecting new livestock owners.

If you are in the United States, you can look at the U.S. Department of Agriculture Farm Service Agency's Hay Net

Web site. (Go to www.fsa.usda.gov/FSA/ and search for "hay net.")

Although it will likely be a more expensive option, you can always go to local feed stores, which typically carry bagged hay. The packages are usually around 40 lbs. / 18.14 kg and contain squares of compressed, chopped hay.

If you live in an area that is experiencing drought, hay is often brought in from other states and may be of greatly inferior quality and full of debris. Under these circumstances, bagged hay is a much better option or you can try an alternative feed.

Alternative Feed

Haylage, if managed properly, is a good fodder. The product is semi-wilted grass or alfalfa that has been dried 55-65% rather than the usually 82-85% drying seen in conventional hay. The fodder is then compressed and sealed in plastic.

In the United States, the major manufacturer of haylage is Chaffhaye (www.chafehaye.com) and there are numerous sources in the UK where the feed is used widely for horses. (Note that many different types of hay can be used to make haylage.)

Although not as readily available, hay pellets are also an option. Llamas and alpacas tend to wolf down pellets, which raises a danger of choking.

It's best to soak the pellets for an hour before feeding, which turns them into a sort of mash, or to put rocks roughly the size of your fist in the feed box, forcing the animals to look for the pellets and thus eat more slowly.

Hay Feeders

The cost of hay feeders varies by the type and use. If you are buying a two-sided covered feeder like those used with sheep that is suitable for placement in a pen you will pay $500 / £298 and up.

An open feeder designed to be attached to the interior wall of a shed where the hay will not be exposed to the elements retails for around $250 / £149.

Both of these feeders hold loose hay. Flat bale feeders for use in the pasture that will hold a rectangular 40 lb. / 18 kg. bale of hay start at $350 / £208. Uncovered round bale feeders cost approximately $175 / £105; covered models start at around $350 / £208.

It is quite common for livestock owners to adopt a "do it yourself" approach to hay feeders, especially if they have a friend who is handy with a welding torch. Use the image browsing section of your favorite search engine to get design ideas.

Managing Stored Hay

When buying more than a bale or two of hay, be sure to work out clear specifics for the purchase with the seller and to get those details in writing. These should include, but not necessarily be limited to:

- price
- terms of delivery
- if additional hay will be held for you
- who is responsible if the hay is sub-standard

When the hay comes into your possession, stack the bales off the ground on some kind of riser like wood pallets and keep it under cover so it's not soiled by birds and other animals.

In many parts of the United States the old-fashioned rectangular bales of hay that weigh 45-100 lbs. / 20.4-45.35 kg are becoming much more difficult to find.

For producers, the bigger round bales are easier, but they range in size from 400-1,200 lbs. / 181.43-544.31 kg. These can be used with llamas, but with several special considerations:

- Make sure that the layers deep inside the round bale are not moldy.
- Remove the strings of baling twine, which are long and dangerous to your animals.
- Use a bale feeder to prevent wastage, but check it daily to make sure the welds are intact so no sharp protrusions will wound your llamas.
- If you have crias (young llamas), they may set up camp inside the feeder, which will necessitate its removal.

Only use the bale outdoors if you have enough animals to eat the entire bale in 3-4 days. When exposed to the elements, hay goes bad quickly. Instead, feed the bale indoors or under cover, unwinding just what you need to give the animals per day.

In areas with heavy rain or snowfall, big bales kept under cover are a better value and will eliminate about 35% of the waste of outdoor use.

If you are storing round bales outside:

- use some means to get the bales off the ground
- cover only fully cured hay with tarps
- don't store the bales close together
- do place them in a straight line
- don't stack the bales

Storing the hay in a rectangular arrangement or stacking the bales will cause the hay to go bad much more quickly.

Grazing

Llamas are happiest free grazing and will thrive when left to roam and graze on grass rather than spending all their time in a corral or barn.

Pastures must be well maintained, however, and should be meadows of fertilized native grasses. You don't want a lot of weeds and thistles in the area, but at the same time lush pastures full of legume hay are far too rich for a llama's metabolism. Animals allowed to graze under such circumstances will quickly become overweight.

You can't just turn your llamas out in the pasture and forget about them. Although they won't challenge fences, the pasture must be secured against predators. The animals need suitable shelter, and lots of clean water. Pastured llamas still need careful husbandry.

Plants Poisonous / Toxic to Llamas

All of the following plants are either poisonous or toxic to llamas and should be eliminated from your pastures.

Aconite	Golden Chain Tree	Peach
Alsike Clover	Golden Rain	Philodendron
Arrow Grass	Greasewood	Poinsettia
Azalea	Groundsel	Poison Hemlock
Baneberry	Heaths	Potato
Barberry	Hellebore	Pothos
Belladonna	Helmet Flower	Privet
Bittersweet	Holly	Ragwort
Bleeding Heart	Honeysuckle	Rhododendron
Boxwood	Horse Chestnut	Rosary Pea
Bracken fern	Horsetail	Rhubarb
Burning Bush	Hydrangea	Rusty-Leaf
Buttercup	Indian Poke	Scotch Broom
Castor Bean	Jack-in-the-Pulpit	Skunk Cabbage
Cherry	Jerusalem Cherry	Sneezeweed
Choke Cherry	Jimson Weed	Soldier's Cap
Christmas	Labrador Tea	Sorrel
Cherry	Laburnum	Sour Dock
Christmas Rose	Lantana	Spindletree
Cowslip	Larkspur	Spurge Laurel

Crown-of-Thorns	Lily-of-the-Valley	Swamp-Laurel
Daphne	Bush	Sweet Pea
Deadly	Lobelia	Tansy
Nightshade	Locoweed	Tansy Ragwort
Death Camas	Lupine	Thorn Apple
Delphinium	Mandrake	Timber Milk-Vetch
Devil's Ivy	Marsh Marigold	Tomato
Devil's Weed	Mayapple	Water Hemlock
Doll's-eyes	Mistletoe	White Baneberry
Dumb-Cane	Monkshood	White-flowered
Elderberry	Moonseed	Rhododendron
Elephant Ears	Morning-Glory	White Hellebore
English Ivy	Mountain Laurel	Wisteria
False Hellebore	Nightshade	Wolfsbane
Foxglove	Oak	Yew
Friar's Cap	Oleander	

Required Shelters

A typical three-sided livestock shelter with one open side away from the prevailing wind will work very well for llamas. Just make sure that the structure will give the animals protection against cold winds and rain while still affording easy access.

The exact arrangement is obviously dependent on region and climate. As a temporary expedient, you can even stack straw bales and cover the top with a tarp. Long-term, however, the animals need a permanent shed.

It is not unusual to see barns with heated floors and automatic misting and watering systems. This is especially true at larger farms with active breeding programs. The truly crucial consideration, however, is space.

Typically with animals that are used to one another from being in the same pasture, 5-7 adults can easily share a 12 x 16 foot / 3.65 x 4.87 meter shed while six females with crias will need a 16 x 16 foot / 4.87 x 4.87 meter space.

Obviously concrete floors with drains are an advantage for cleaning and washing out urine. If the floor of the shelter is dirt and crushed stone or gravel, the material should be removed and replaced periodically.

For the most part bedding is not necessary. Llamas have thick coats that will keep them warm. If bedding is used, always choose straw, but muck out and clean the material daily as you would with horses.

Clearly it's impossible to offer a "ballpark" estimate of construction costs. If you are a "do it yourself" type, you may be out nothing more than materials and time, but if you have to hire a contractor, the cost of labor will be included in the total price of the project.

Fencing and Fence Materials

The primary goal of fencing your llamas is to protect them from predators. The proper fencing needs to be in place BEFORE you buy your animals. About one acre of land will carry 3-5 llamas, so plan your pasture area appropriately.

Adult llamas don't challenge a fence, but young animals will crawl under plank and post fences unless the fence is also lined with woven wire. You'll need to pick your materials carefully.

Depending on the number of animals you're keeping, it's ideal to have sufficient land to rotate the animals through multiple pasture areas. This will prevent over-grazing and allow the available forage to recuperate during rest periods.

Avoid all fences with gaps into which llamas can stick their heads. Traditional cattle fencing (wire net at the bottom with strands of wire on top) is not appropriate.

Some favored fencing types include welded mesh "no climb" horse fencing and chain link or cyclone fences. Don't use single strand high tensile "New Zealand" fences, barbed wire, or plastic fencing.

The most secure and safe form of lama fence is installed woven wire (sometimes called field fence or field mesh.) The wire is made of smooth horizontal strands held apart by vertical stays. The material is sold in three versions:

galvanized, high-tensile, and polymer-coated high-tensile.

Each variation has the vertical strands positioned at either 6 or 12 inches (15.24 cm to 30.48 cm) and are made in heights from 26 to 52 inches (66.04 cm to 132.08 cm).

Avoid any wire openings that are more than 6 inches square (15.24 cm). Posts should be placed at 14-16 foot / 4.26-4.87 meter intervals to support the wire.

You can tell what kind of woven wire you're buying by looking at the numbers on the roll. If it's marked 8/32/9, you're looking at wire with eight horizontal wires that is 32 inches / 81.28 cm tall with vertical stays at nine-inch / 22.86 cm intervals.

Although more expensive initially, and time-consuming to install, woven wire fences are attractive, safe, and require little in the way of maintenance.

If possible, have double fences. The outer fence for controlling predators should be 5-6 feet / 1.5-1.8 meters high. Bury mesh wire at the base at a depth of 1 foot / 0.3 meters. Electric fencing is also an option for the perimeter, but not to come into contact with your llamas.

Regardless of type, the fencing to contain the llamas should be 4-5 feet / 1.2-1.5 meters high.

Check both perimeter and interior fences daily and replace all damaged materials or areas where a gap may be forming.

Portable fencing panels can be used when rotating grazing areas, subdividing a pasture, or making small areas for females with cria or making a quarantine pen.

As is the case with shelters, fencing costs are difficult to estimate. They depend on area to be enclosed, material chosen, and the cost of labor. It's always a good idea to get multiple estimates, and, if possible, to see an example of a fence your contractor has built for someone else.

Additional Equipment

As with any kind of livestock, it always seems there's "one more thing" you need to take care of your llamas. Over time you'll acquire all the little stuff from an assortment of buckets to brooms, rakes, shovels, wheelbarrows, and the like.

There are, however, some items you will definitely need in your standard store of equipment from the beginning.

Halters and Lead Ropes

It is imperative that llamas be trained to haltering even if they must be placed in a catchpen or stall. Do not leave a halter on a llama full time, however.

The halter straps will cause calluses, ulcers, and abscesses, and can even break the animal's neck if the halter becomes caught on something and the llama panics while trying to free itself.

NEVER leave the same halter on a growing llama! The flesh will grow around the straps, particularly at the noseband. Beyond being painful, this can also cause the nasal passages to become malformed.

Make sure that all halters are properly fitted:

- The noseband should sit 1-1.5 inches / 2.54-3.81 cm above the end of the nose bone. If it sits lower, it will obstruct the nose and cut off the airway.

- You should be able to place your hand between the underside of the jaw and the bottom halter ring. Halters that are too tight will cause sores to develop.

Use only halters that have been specifically designed for llamas. Each halter should cost less than $20 / £12. Any livestock lead rope with a good catch on the end will work well with llamas and should cost no more than a halter.

Shearing and Shearing Equipment

Shearing your llamas is primarily a necessity for health and comfort, but the fiber is valuable and can be harvested for profit. Regardless, you will need to learn how to shear your animals.

Some people prefer to hire professionals to do this task for fear of hurting the llamas, however, your money might be better spent paying someone to teach you how to do the shearing yourself in the future since this is an ongoing husbandry chore.

Use a blower (a leaf blower will work perfectly well) to get debris off the surface of the fiber. Keep the nozzle far enough away from the animal that the blast of air causes the fiber to fan out.

Don't try to brush or remove tangles from the fiber prior to shearing. It's painful for the animal and you won't make any progress.

Plan to leave approximately one inch of fiber on the animal to keep the llama from developing painful sunburn. Typically the animals are placed flat on their sides and the fleece is removed starting at the top of the back.

Work down, making parallel cuts. Talk to the llama as you clip. Even if the animal is perfectly still and behaving beautifully, it's scared.

Llamas have good memories so don't lose your temper. The animal will react with even more stress the next time your shear, associating the activity with a bad reaction on your part. Llamas tend to settle down after 10-15 minutes no matter what they're displeased about, so remain patient.

I recommend going on YouTube and watching videos of llamas being sheared to have a better sense for the overall process. The clipping is always done in a closed shearing barn so the animals can be controlled more effectively.

Understanding Your Tools

You can use hand clippers, but in most cases power shears are the preferred tool. The shears require some practice, but they make the clipping go much more quickly and efficiently.

Power shears are comprised of a hand piece, comb, and cutters. The best option is to get a unit with the motor built into the handle so you're not fighting a long power cord.

Good quality commercial models cost $250-$500 / £150-£298. The comb, which attached to the hand piece, is adjustable.

The flat side faces up and away from the animal so it can enter and separate the fiber. You will find that this piece dulls quickly and won't be good for use on more than 2-3 animals before it must be replaced. Additional combs cost $15-$35 / £9-£21 each.

Cutters have four triangular points. They are also attached to the hand piece and fit snug against the comb. The blades of the cutters dull even more quickly than the combs. Expect to use three cutters per single comb at a replacement cost of $10-$15 / £6-£9 each.

A Word on "Cuts"

As you work, avoid making "second cuts" which are short cuts made in the interest of improving the appearance of the animal after the clip. If you're marketing your fiber, the second cuts are worth much less and people who do hand spinning absolutely hate them.

Considering second cuts is important, since the individual artisan market for llama fiber is a good way to sell the material for profit if you only keep a few animals. The higher the quality of the fleece and the more carefully it has been sheared, the better the price it will command.

If you're only clipping the llama for aesthetics and health reasons, remove the prime fiber, which you can still sell, then sweep the area and go back and touch up the animal's "hairdo." Remember the standard rule of all haircuts regardless of species, "It will grow out."

The finest and best fiber comes from the body, and should be separated and distinguished from the wool on the neck and leg. Clean grain bags or old pillowcases are good to store fiber, but never use plastic. Be sure to keep the wool where mice and moths can't get into it.

Toenail Trimmers

Toenail trimming is a definite part of llama husbandry, but the frequency with which you will need to perform this chore will vary from animal to animal.

If your llamas are kept on rocky soil, they may keep their nails worn down, but those that live on soft ground could need a "pedicure" every six weeks.

Unfortunately, llamas are prey animals and they are strongly resistant to having their feet picked up. Instinct tells them that if they can't break and run, it's game over.

To complicate things even more, males fight and play by biting at each other's lower legs, so they naturally interpret any movement toward their extremities as something to be resisted.

Often a handle chute will be necessary to successful complete a nail trim. Don't try this procedure on your own without some hands-on instruction from a knowledgeable party, and always wear protective clothing.

Llamas don't spit on humans often, but when you mess with their feet, all rules of polite behavior go out the window.

Put your hand on the animal's shoulder or haunch and run it down the leg until you reach the foot. Choose a verbal command that you will use consistently like "foot" or "up." Grasp the llama's foot firmly and pick it up without hesitation, but don't be aggressive. You want to convey confident control, not dominance.

Practice with your llama before you actually intend to trim the animal's nails. Always reward the llama with a treat for picking up its foot. These animals have good memories and will hold on to negative associations forever! Move slowly and let the llama get completely comfortable.

When you are able to pick up the llama's feet, the actual clipping process should be quick and smooth. Again, convey confidence. The llama will pick up on any signs of fear or stress on your part. Don't allow your nervousness to agitate your animal.

- Cradle the foot in your hand so that the underside is up and clearly visible.

- Use a pair of garden pruning shears.*

- Carefully trim the nails until they sit level with the bottom of the foot pad.

*There are many kind of clippers that will work well to trim your llama's toenails, including those specifically designed to be used with sheep and goats. Be sure that the tool you select fits your hand well, is easy to use, and is sharp enough to get the job done.

When you have clippers that feel comfortable to you and are well sharpened, you'll have maximum control so you can work quickly and efficiently. The llama will be much happier with the whole business if you get it over with fast and with a minimum of fuss.

Vaccinations

Required annual vaccinations for llamas will vary by region, but a good starter program will include the following annual shots, typically given in the spring.

If your vet has not treated llamas in the past, consult with other owners in the area, or with your local agricultural extension service to determine if any additional inoculation is needed.

- **C&D Toxoid** – Prevents over eating disease caused by *clostridium perfringens* that multiply and produce deadly toxins in the digestive tract.

- **Tetanus Toxoid** – Prevents "lock jaw" caused by *clostridium tetani* introduced through cuts, puncture wounds or any accidental lacerations.

- **Rabies** - To prevent rabies caused by bites from infected animals including, but not limited to, skunks, foxes, coyotes or bats.

Most vaccines are given initially in two doses spaced 30 days apart. Young animals receive their first shots at three months, with all vaccines "boostered" on a yearly basis.

Meningeal worms, which are common intestinal parasites in whitetail deer, are life threatening to llamas. The parasites are passed through snails and slugs the llamas ingest while foraging.

Depending on your area, and the advice of your veterinarian, monthly injections or oral doses of Ivermectin may be required to protect your llamas from these potentially deadly pests.

Chapter 4 – Understanding Basic Llama Health

Although llamas are hardy animals with a low maintenance profile, they do have specific health requirements. If you don't know, for instance, that not all llamas can lick easily due to their attached tongues, you may think a regular livestock mineral block is appropriate for these South American camelids. Some llamas however, do better with granular mineral supplements.

In some areas where good hay isn't available, grain should be used in a llama's diet, but because their digestive system involves active fermentation, too much grain can cause gastrointestinal upset and bloat.

I do not pretend to be able to give you an absolute definitive guide to llama health. The following chapter highlights some of the unique characteristics of a llama's anatomy, and discusses common problems in the species.

To truly safeguard the welfare of your animals and to know how to care for them relative to your location and climate, you need a veterinarian who has experience with llamas or a mentor in business who can offer guidance while you're learning how to work with your animals.

Potential for Nutritional Deficiencies

Although anyone can learn the nutritional basics llamas require, the correct forage and necessary supplements will vary by region based on factors like climate and soil type. Especially if there are no other llama owners in your community, speak with a livestock nutritionist or agricultural extension agent to make good choices for your animals.

Do not, however, make any abrupt changes in your llamas' diet. The animals can easily develop severe digestive upsets that are potentially fatal. Llamas thrive on routine and don't react well to the stress caused by missed or delayed meals. It's important to observe your animals closely and make sure they are eating. Llamas that "go off their feed" are usually sick.

(Please see the relevant sections of Chapter 3 for a better understanding of llama nutrition in regard to hay, grains, and specific supplements.)

The Specialized Llama Digestive System

As modified ruminants, llamas have forestomachs with three compartments rather than the four compartments seen in true ruminants like sheep, goats, cattle, and deer.

The three sections of the forestomach are C1, C2, and C3, each with a highly specialized function. The largest digestive compartment, C1, is on the left side and comprises 80% of the total stomach volume.

The C1 chamber does not secrete enzymes, but is basically a fermentation vat filled with microoogranisms that convert cellulose into digestible nutrients.

Food enters the C1 chamber through the esophagus so fermentation can begin. Coarser pieces of food are regurgitated in the form of a bolus or cud, which is chewed and swallowed again.

Each time the food is chewed, the additional saliva adds bicarbonate and phosphate buffers to the mixture. This helps to stop the formation of acid during fermentation. Llamas in good health will ruminate for about 8 hours during every 24-hour period.

Only the true stomach is functional in newborn llamas (crias). The baby derives the necessary microbes to start the function of the C1 chamber from nursing its mother. By 8 weeks the C1 has reached its full adult size and by 12 weeks the entire digestive system is functional.

Llamas are highly efficient rough pasture feeders, consuming 20-40% less feed per unit of metabolic weight than sheep, in part because they produce more saliva in relationship to the volume of the foregut.

The basic elements of the llama diet are pasture, hay, concentrates, minerals, and water. The best diet is based on high-quality pasture and hay, with long fiber grass hay for dry forage. Avoid feeding high-protein hays like alfalfa, clover, lespedeza, and other legumes in any great quantity.

Hydration is Essential

Fresh, clean, cool water is necessary for llamas in the summertime and they must have access to thawed water when the temperature falls below freezing.

Use multiple watering troughs to prevent territorial llamas from staking claim to their favorite and depriving other animals of access to the water.

Select troughs that sit 12 to 15 inches / 30.48 to 38.1 cm off the ground and that will hold approximately 50 gallons / 189.27 liters.

If you can afford automatic waters that refill when the level drops to a predetermined volume, use them, otherwise you'll need to check and refill your troughs daily – or more often in warm weather.

Common Health Issues

The following health problems are some of the more common issues that arise with llamas kept for all purposes. Some can be corrected by changes in basic husbandry practices, but all require the services and knowledge of a qualified veterinary professional.

Angular Limb Deformity

During the fall and winter months, llama crias may be born with Angular Limb Deformity (ALD) because the decreased levels of sunlight prevent adequate processing of Vitamin D. The condition, however, may be congenital or acquired.

Often the underlying cause of ALD is abnormally rapid growth at the growth plate or physis on one side of the bone only.

The growth plates are located at the end of the long bones, so a difference in growth rate can change the angle of the limb distally (away from the center of the body.) This is analogous to the state of being "knock kneed" in humans.

If the cria is born with ALD, the condition may be a consequence of factors other than Vitamin D deficiency in the mother including intrauterine malposition, premature birth, and trauma at birth, or abnormalities of the tendons and ligaments.

ALD can also develop in instances of asymmetric weight bearing from trauma, infection, or joint inflammation.

Although visual diagnosis is common, the amount of fiber on the limb often prevents a precise measurement of the angle of the deformity. Then it becomes necessary to wet the fiber, tape the limb, or shear away the fiber to obtain a clearer view.

X-rays are taken to pinpoint the origin of the defect, the angle of deviation, and to locate and identify any other adjacent problems like swelling in the soft tissues.

Age, severity of deviation, and the intended use of the animal dictate the course of treatment. In llamas with insufficient growth and development of the structures supporting the bones, splints are often applied for 7-14 days with Vitamin D supplementation administered.

This approach is typically taken with cria younger than 3 months. If the condition is not detected until after 6 months, surgical intervention may be necessary.

This is only an option in animals with actively growing bones and includes procedures to stimulate bone growth, to implant screws and metal orthopedic wires for straightening, or to remove a wedge of bone to redirect growth.

Bovine Viral Diarrhea (BVD)

Increasingly llama owners should be concerned about the bovine viral diarrhea virus, which is one of several pestiviruses known to infect domestic and wild ruminants, swine, and camelids around the world.

When BVD is present in cattle, the health consequences include decreased weight, poor milk production, aborted births, and death. Like any virus, BVD can present in a variety of ways from a mild infection to diarrhea, respiratory distress, and hemorrhage.

Typically, adult llamas have only a light case of BVD, but if a pregnant female contracts the virus, she will likely lose her baby or give birth to a cria suffering from congenital birth defects.

In some instances, cria can also be born with a persistent BVD infection. The animal will then shed the virus throughout its life as a carrier, which can have devastating consequences for a herd. A true understanding of the effect of BVD in llamas is still evolving and unfortunately it is not even clear by what vector the virus is transmitted to camelids.

There is no approved BVD virus for use with llamas, but there are preventive measures that can help and should be implemented in any area where BVD is known to be present. These include:

- Keeping a closed herd
- Maintaining strict biosecurity for incoming animals (quarantine)
- Periodic screenings of open herds

Internal Parasites

In general, llamas can be infested with any of the internal parasites present in sheep, goats, and cattle, but are less plagued by external parasites due to the density of their fiber.

Completely eradicating internal parasites is impossible with livestock, especially if your land carries multiple species; therefore part of ongoing health care for your animals includes controlling "worms."

The traditional method for dealing with worms in any species of livestock has always been to "drench" all the animals at the same time of year with a rotating pool of oral deworming products.

The strategy was believed to avoid instances in which the parasites developed immunity to the curative agents. Unfortunately, this did not prove to be the case, and over the last couple of decades many worm species have become highly resistant to deworming medications.

For this reason, it's best to have your llamas screened for internal parasites by your veterinarian on a regular basis and to follow the vet's recommendation for specific dewormers. Otherwise, you may simply be stressing your llamas by giving them medications that do them no good whatsoever.

Heat Stress

It's important to remember that llamas are indigenous to the high Andes Mountains of South America where temperatures rarely exceed 75 F / 24 C. Llamas can easily suffer fatal heat stress if subjected to extremes of heat and humidity.

The International Llama Association recommends the following formula for determining how a given climate will affect your animals:

- Take the ambient temperature in degrees Fahrenheit.
- Take the humidity.
- Add the two.
- If the number is less than 120, your llamas are not at risk.
- If the number is greater than 150, take as many precautions as possible.
- If the number is approaching 180, the risk of heat stress is extreme.

Be especially vigilant with animals that are:

- Very old

- Very young
- Ill
- High-strung / nervous
- Obese

Symptoms of heat stress include:

- Panting
- A depressed attitude
- Failure to eat
- A rectal temperature of 140 F / 40 C
- A heart rate in excess of 90 beats per minute
- Drooping lower lip
- Paralysis of the face
- Slobbering
- Overall body weakness
- Trembling
- Scrotal swelling (intact males)
- Seizure

By the time a llama reaches the point of collapse and seizure, the heat stroke will likely prove fatal no matter what means of intervention you attempt.

If you suspect heat stroke, put the lama in the shade, in front of fans, or inside an air-conditioned room. Take the llama's temperature and record the reading. Repeat in an hour to judge the success of your efforts.

Do not hose the animal down unless it has been sheared. Only wet the underbelly and rectal area with cold water for 15-20 minutes per session.

ALWAYS shear your llamas when warm weather starts and provide them with shelter, including cold, clean water and potentially fans.

Brucellosis

Although rare, brucellosis can occur in llamas and a blood test for the disease is required for entry of the animals into some states in the United States. The test does often result in a false positive, however, therefore a second test is considered standard procedure for verification.

Brucellosis is transmitted by bacteria and causes fever, stiff and swollen joints, and spontaneous abortions with retained placentas.

Symptoms include spontaneous abortions, retained placentas; intermittent fevers; and stiff, swollen joints. Typically livestock regulations require that animals with brucellosis be destroyed. The best prevention is the brucellosis vaccine, but it is an off-label usage in llamas.

Caseous Lymphadentis

The bacterium *Corynebacterium pseudotuberculosis* causes *Caseous lymphadenitis* (CL) or "cheesy gland." While not a common condition in llamas, any lumps on your animals should be evaluated.

The lumps present in CL are thick-walled abscesses that are cool to the touch and filled with a greenish-white, odorless pus with the consistency of cheese.

The masses form on lymph nodes and in lymph tissue on the flanks, chest, and neck. If present internally, CL abscesses can be found in the abdominal cavity, on the lungs, liver, kidneys, spleen, and brain as well as the spinal cord.

The bacteria is transmitted through the pus when an abscess ruptures, therefore any llama with such a lesion should be quarantined. The abscess should be drained and treated according to instructions from a veterinarian. Care must be taken with the drained pus to prevent further contamination.

Please note that CL can be transmitted to humans, so protective clothing should be worn and sterilized or burned after use along with any other materials contaminated with pus.

Botflies

Llamas, sheep, and goats are all susceptible to infestations of nose bots, a yellowish, hairy fly about the same size as a common housefly. These insects, which are often mistaken for bees, lay their first-stage larvae in the nostrils of livestock.

One female botfly can lay as many as 500 eggs that migrate up the nasal passage when they hatch and feed on mucus and the mucous membrane. The affected animals begin to shake their heads. They lose their appetites and suffer from a constant opaque discharge from the affected nostril. Currently, the most effective treatment is Ivermectin.

Fighting Teeth

At 18-24 months of age intact male llamas develop six sharp fangs or fighting teeth. There are two on the top gum and two on the bottom on each side of the mouth. Females and even geldings can also have fighting teeth, but not to the same extent. These teeth must be removed or blunted or the males will use them to fight with serious and bloody results.

The extraction process can be risky and may seriously injure the jaw. It's better for the teeth to be sawed off periodically to gum level until they stop growing when the animal reaches full maturity. Vets use a flexible cutting wire or a grinding tool like a Dremel to saw the teeth down to the gum line.

Llamas have no upper teeth. The lower teeth meet the upper dental pad. Sometimes young alpacas or their aged counterparts suffer from dental malocclusions or protruding front teeth that will also need to be trimmed back to facilitate chewing.

Care should be taken not to grind a tooth down to root level, which will expose the animal to infection. Teeth should never be cut with any kind of nipper or cutter because they will shatter.

Euthanasia and Burial

Inevitably as your llamas age, you will be faced with euthanizing animals with medical conditions that cannot be treated, including deterioration from advanced old age. It is always difficult to make such a decision, and the choice carries extra complications with livestock that are purely practical in nature.

With large animals, it's important to have a plan in advance for how you will deal with the remains, especially if you do not have enough land to bury the animal on your own property. If you live within a municipality, you should understand the regulations that may apply. It is highly unlikely that you will be able to bury a llama in your backyard.

There are livestock removal companies that will take away deceased animals for a fee. The cost is typically between $250-$300 / £148-£178. Your veterinarian or other llama owners in the area should be able to recommend a company you can contact.

Cremation is an option and is often chosen in cases where the animal is a beloved family pet. Understand, however, that the cost may be as high as $1000 / £595 for the service and $200 / £119 if you want the ashes back. The higher price is because in instances where the ashes are returned, the cremation is exclusive to your animal's remains only.

When it is time to euthanize your llama, the animal should be moved to an area where the remains can be dealt with easily for burial or removal. Preferably choose an area into which a truck or van can be positioned to carry the body away.

Your veterinarian will inject the animal with an anesthetic, likely penta-barbitol. The dosage is based on the llama's correct body weight and constitutes an overdose. The drug instantly halts brain and heart activity. The llama will feel nothing. Death is instantaneous.

When the procedure is complete the body must be arranged before rigor mortis sets in. Place the back and front legs of the animal in the kush position, securing them in place with ropes. Bring the head and neck forward and down to the side, also using rope to hold them in place. This compact position will facilitate handling and/or burial.

If this responsibility is too much for you, be sure to ask someone to help you. If the body is allowed to stiffen after death, it will be much more disturbing and unpleasant to remove the deceased llama. Positioning and tying the limbs actually allows for much more dignified handling of the remains.

Chapter 5 - Llama Breeding

In designing a llama-breeding program, the intended use of the animals is the primary consideration. If the animals are to be companions, personality and docility are clearly among the most important traits.

Long-term goals should be the guiding principle in selecting breeding stock. The actual "mechanics" of the process are straightforward and typically the animals require little if any assistance.

Reproductive Physiology

Female llamas have no regular estrus cycle. They are induced ovulators. The act of mating causes their system to produce an egg "on demand" for fertilization within 24-36 hours.

"Open" females should not be bred until they have reached 18-24 months of age, although they are capable of conceiving earlier in life. Allowing a female to carry a cria at too young an age, however, can be dangerous for the llama and her unborn offspring.

Although males are fertile by 7-9 months, they aren't dependable sires until at least three years of age. By that time they are both sexually and socially mature. Since males can exhibit a great deal of hormonal aggression, allow the "boys" to be fully mature before bringing them into a breeding program.

Like all South American camelids, male llamas are dribble ejaculators therefore copulation can take as long as 45 minutes. Breeding takes place with the male on top and the female in the kush position.

Prenatal Care and Specific Concerns

In general, it is not difficult to care for a pregnant llama. Many females that work as pack animals continue to do so until the later months of their pregnancy. Again, llamas are very healthy animals by nature, and it isn't necessary to think of a female llama as "delicate" while she carries her baby.

Sound nutrition and stress reduction are the best protections against premature birth in llamas. Any physical assistance to a female during the actual process of giving birth should only be administered when absolutely necessary.

Llama cria should not be "pulled" like other livestock. The baby can be fatally injured if handled incorrectly. It is absolutely essential to seek qualified veterinary assistance for all difficult deliveries.

Delivery of the Cria

Llamas can give birth at any time during the year, although it is best to avoid the temperature extremes of high summer and deep winter. If an unplanned pregnancy occurs so that a birth will occur during one of these unseasonal periods, make sure the female has adequate shelter, even to the point of keeping her confined as her delivery date nears.

Females carry the young for approximately 350 days, giving birth to a single cria. It is very unusual for llamas to have twins, and when such a birth occurs the offspring typically do not survive. Often when more than one embryo forms, it is reabsorbed within the first two months of the pregnancy and the female is then open for a second breeding.

Females give birth during the daylight hours on almost all occasions, typically between the hours of 7 a.m. and 2 p.m. Delivery requires 10-45 minutes from the time the feet and head become visible. In this regard, llamas are much more considerate of their owners than other types of livestock. You will rarely if ever find yourself up at 2 a.m. helping a llama deliver her cria.

Mother llamas do not lick their offspring clean, nor do they consume the afterbirth. The impact of the cria hitting the ground clears the airway and initiates breathing. This is natural and should be allowed to occur without intervention.

At birth newborn llamas weigh 25-30 lbs. / 11.34-13.6 kg on average, but can be anywhere from 18-49 lbs. / 8.16-22.22 kg in size.

The Cria Post-Delivery

After giving birth, the mother llama stands protectively nearby her offspring and hums constantly, but she does not touch the newborn, which will also begin to hum. This behavior is not evidence of neglect, but the beginning of an intricate and intimate means of communication.

Humming helps to guide the cria during these critical first moments and ultimately serves to stimulate the baby's first feeding. Thereafter humming becomes an ongoing form of mother/baby identification.

After thrashing for about 5 minutes, the cria will get its legs drawn under its body and into the kush position and then stands for the first time. If you are watching, don't be tempted to help. The baby knows what it's doing and won't injure itself. Once the cria is standing, the mother moves over her baby so the newborn can find an udder and begin to nurse.

Within four hours of the delivery, the mother passes the placenta. Occasionally a placenta will be retained, extruding partially from the mother's body. Do not attempt to remove it, since doing so may result in hemorrhaging.

Seek the aid of a veterinarian who can safely loosen and remove the tissue.

Cria Nutrition

The first milk a baby llama receives, colostrum, is rich in antibodies and crucial for building immunity. Growing crias nurse frequently immediately after birth, but less as they age. Their mother's milk is rich and densely concentrated with nutrients, but is not produced in large quantities.

Following birth, it may be necessary to supplement the cria's diet with Vitamin D to ensure proper bone formation and growth. This can be accomplished either with oral supplements every 2-4 weeks or with an injection given every 60 days.

If the birth occurred in the spring, the cria will be less likely to require Vitamin D since they will be able to get an adequate supply from sunlight. Babies born in the fall or winter, or those that are contained indoors for long periods will require supplementation.

By six months of age, young llamas are able to forage for themselves and the mothers become much less protective. Typically supplemental nursing is not required after this stage of life. A young llama should be matured fully by 2-4 years of age.

Re-Breeding Females

Normally females are bred back 3-4 weeks after giving birth. A blood test to detect progesterone is used to confirm pregnancy 21 days after breeding. The female's behavior is also a reliable indicator of conception. Pregnant llamas will refuse the advances of herdsires to which they are introduced.

Males that will not be used for breeding should be gelded by three years of age, but not before 18 months. Geld bottle fed males sooner to prevent aggressive behavior. Also, since llamas, guanacos, alpacas and vicunas are capable of interbreeding, keep the three species in separate pastures.

Selecting Herdsires

Selecting good herdsires is important for the future of any breeding program, but is not as easy as you might think.

Obviously major considerations in picking a sire include such things as:

- Conformation
- Fiber quality and color
- Overall appearance
- "Presence" and disposition
- Pedigree

The most important thing, however, is simply fertility. If a male does not have good reproductive qualities, he is clearly of no use in a breeding program. There are, however, no set parameters to evaluate male fertility as there are with other species.

Most large breeding operations handle this problem by keeping meticulous performance records. Smaller operations and individuals generally take their chances, or contract out to stud farms for the services of their herdsires.

Depending on the pedigree of the animal in question, stud fees can be as little as $50 / £ 30 or as much as $1500+ / $885.

Chapter 6 – Llamas as Pack Animals

Llamas are excellent pack animals for hikers because they can carry a substantial amount of gear, are easy to take care of, and have a minimal environmental impact.

It is important, however, to train a llama well for its work as a beast of burden, and this is not as straightforward as simply throwing a pack up on his back.

Male or Female?

Traditionally male llamas have been used for packing due to their sturdier frames and slightly larger size. In truth, however, either gender can make excellent packers if handled appropriately.

One argument has always been that females are not physically and emotionally capable of packing and also delivering healthy cria. With correct training, however, neither physical conditioning nor potential psychological stresses are an issue.

Even pregnant llamas can serve as pack animals with owners who understand the hormonal changes their bodies will experience and how that correlates with their likely behavior.

Female llamas may need planned "breaks" from work, but they are just as capable of serving as pack animals in their first few months of pregnancy as a mare would be.

Teaching Llamas to Work as Pack Animals

Have the following items on hand. Never try to train any kind of animal without all of the materials you need within easy reach:

- halter
- wool blanket (twin or full size)
- training pole or wand
- bath towel, dishtowel, and handkerchief
- 2 lead ropes
- assorted old rags

You'll have more success if you work with the animal in a 10' x 10' / 3.04 x 3.04 meter catch pen. Tying the llama up will only reinforce his feeling that something threatening is occurring.

Use a lead rope on the llama to give you some control as you work, but allow the animal to move freely in the pen. The rope should be used only to indicate direction, don't try to slow the llama down or get him to stop.

Understanding the Animal's Perception

You always want to start with an animal that is easy to halter and to lead, and that is receptive to having his legs and feet touched. With a solid candidate for pack work, you can follow a series of simple steps to get the llama accustomed to his new "duties."

The best pack animals are those that have been exposed at a young age to all the things that might frighten them on a hike in the woods. Llamas are not necessarily born packers, but they can greatly enjoy the experience when properly trained.

You must never, however, lose sight of the fact that llamas are prey animals. They spend their lives interpreting and evaluating the signals they receive from their surroundings for perceived threat potential. A llama has no instinctually positive frame of reference for the physical sensations a pack creates.

It just knows "something" is clutching it around the middle and won't let go. The very idea screams "predator" in a llama's mind, with panic being a perfectly reasonable – and potentially life-saving – response so far as the animal is concerned. Understandably, introducing a llama to a pack in this way creates long-term avoidance behaviors.

As with any animal, if your llama clearly becomes bored, agitated, or aggressive, halt the lesson and start again the next day. Llamas can be very stubborn by nature, but they also assimilate information extremely well. You can usually teach them tasks in steps, picking up one day basically where you left off the day before.

Do not mistake the "freeze response" for cooperation and acceptance on the llama's part. This is another natural response to fright. All may seem to be going well as you position and cinch the pack, and then the llama snaps awake and basically loses his mind in an attempt to also lose the pack.

Giving the animal freedom to walk around the pen as he becomes familiar with being "saddled" should help to avoid this reaction. Offering a bite or two of food from time to time during these lessons will also facilitate a positive response and remind the poor llama to just breathe!

Camelids have a tendency, like many humans, to hold their breath when they're afraid, which increases the chance of erratic and even explosive responses.

Introducing the "Pack"

As you begin to introduce your llama to a real pack for the first time, take the following steps:

1. Herd the llama into the catch pen.
2. Halter the animal.
3. Put a towel folded in quarters on the llama's back.

Let the llama walk around with the towel in place for a couple of seconds and then drag the towel off its back, letting the material touch the animal's legs as it drops down.

This last step is to help the llama get used to anything that falls at its feet in the future. Don't be surprised if the llama gets upset enough to kick at the material.

It may be necessary to pick a smaller piece of material, like the handkerchief and work up progressively to something bigger. Don't move up to a larger item until the animal is used to the smaller one.

Remember that llamas are very curious by nature. Let the animal check out everything you put on its back. Once the llama is comfortable with the cloth folded in quarters, switch to a half fold and start over. Keep working until the material is completely unfolded

If possible, graduate up to a blanket that is long enough to almost touch the ground when it's fully unfolded. This will further accustom the llama to having something on his back and around his legs. When you've reached this point:

1. Fold the blanket in fourths.
2. Put it on the llama's back.
3. Drape a lead rope over the animal's back.

Stand close to the animal by the llama's front shoulder facing its rump. Reach over the llama's back with one arm and push the rope into your other hand. Think of the motion as if you were planning to give the llama a hug around his midsection. Move slowly, and do nothing to startle the animal.

You are creating a practice cinch with the rope and allowing the llama to get used to the sensation of pressure on its back and belly simultaneously. When you stand, use your hands and arms to press down gently on the llama's back, and then with an end of the rope in each hand, lift up gently in the cinch area and release. If all goes well, tie the blanket snuggley on the animal.

The vast majority of packs are designed with two cinches, so pass a second lead rope over and under the animal, positioning the rope at the base of the ribs and, in male animals, about the width of a hand in front of the sheath.

When the llama will comfortably walk around the catch pen with the blanket in place, snug up the ropes and walk the animal for a brief period. Let the animal get used to the feel of the gear while walking and trotting and as he brushes up against things. Return to the pen.

Next graduate to several folded towels or even an old pillow tied to the cinch to simulate panniers. By the time the animal is comfortable with this arrangement, a real pack will not feel substantially different to him and should not elicit a panic response.

Introducing a Real Pack

When you introduce the llama to a real pack for the first time, always give the animal a chance to check out the equipment before you put it on his back. Work in a catch pen and follow all the same steps you used in training, including pulling the pack off a couple of times to both the right and the left before you attach the cinches.

Most experts are in agreement that even though "training packs" are available for this purpose, they are not as effective as the "do it yourself" method described here. By design, training packs are typically of one-piece construction and don't allow enough options for getting the llama gradually used to all the sensations of being saddled.

Teaching the Llama to Tie

Obviously there is very little involved on your part to tying up an animal, but it is crucial that the llama accept this arrangement peacefully. If a tied llama fights against the lead, even one with elasticity, serious muscle and spinal damage can result.

    1. Work inside a stall or catch pen.
    2. Halter the animal.
    3. Use a long, flat lead.

Pass the end of a lead one time around a fence post or through a tie ring, holding the loose end in your hand. The line should be able to slide and release if the llama pulls against it.

Stand quietly and wait for the llama to back up. When he does, hold the line and offer light resistance. If the llama starts to struggle, tap him on the back leg with the training pole or wand, using it to gently encourage him to move forward and away from the pressure he is feeling. If he begins to panic, release the resistance and let the llama calm down.

Speak to the animal in a steady, comforting voice until you are ready to begin again. This time you may also want to flap a towel when the llama backs up, still using the wand to teach him to move forward. With this exercise, you are teaching the animal both to relieve the pressure and to stop a scary "thing" from happening. Do not progress to the next step until you're able to use a longer line and move some distance away from your llama with the animal standing calmly.

The first time you actually tie the llama, use a quick release knot and stay close by. You can also use a material like baling twine that will snap and come loose if the animal truly panics and begins to struggle.

First Outings

I would recommend trying your llama on short outings with a light pack for the first two or three times you take an excursion together.

The animal will quickly become accustomed to having the sensation of weight on its back, and generally a llama's native curiosity kicks in and he's more interested in the world around him than what he's carrying for you.

Llamas can carry 25-30% of their body weight for several miles, so they are excellent and efficient hiking companions.

It's a good idea to make sure your llama has had a nice meal of hay before you take to the trail to lessen their natural behavior of trying to forage while you're trying to hike!

If you're staying out overnight, pick a grassy area where the animal can safely graze and take along some pellet food or chopped hay.

Always hike with plenty of water for yourself and your llamas.

Pack Llama Trial Association (PLTA)

In the United States, the Pack Llama Trial Association is a non-profit corporation that was created in 1998 as a way to competitively test llama's pack competency against a written field trial standard.

(The organization does not maintain a website per se, but has a Facebook presence. Search on Facebook for "Pack Llama Trial Association.")

Many professional packing companies participate in these trials to gain PLTA certification as a way to document the ability of their animals. The trials focus on real-life packing conditions rather than the performance of any "tricks," therefore they are a realistic assessment of a llama's natural abilities and the quality of its training.

There are four levels of certification:

- **Basic** demonstrates mastery of minimal packing requirements.
- **Advanced** requires more rigorous training and qualification for "moderate" packing requirements.
- **Master** level animals are trained for serious packing.

- **Extreme** certification is for true llama athletes with superior and proven skills.

It's generally assumed that mastery level certification represents approximately four years of training and conditioning.

Where Can I Go with My Llama?

Obviously you are always free to use llamas when hiking on private land with the permission of the owner. In the United States, llamas are used widely in the national park system due to their minimal environmental impact. The applicable National Park System regulation reads in part:

> *8.2.2.8 Recreational Pack and Saddle Stock Use*
> Equine species such as horses, mules, donkeys and burros, and other types of animals (including llamas, alpacas, goats, oxen, dogs and reindeer) may be employed when it is an appropriate use to support backcountry transport of people and materials and will not result in unacceptable impacts. NPS regulations at 36 CFR 2.16 prohibit the use of animals other than those designated as "pack animals" for transporting equipment.

The safest and most reliable option in planning a trip with your pack llama is to check with the specific national (or state) park you wish to visit to find out if and how you may hike with your llamas and what permits are required.

In the United Kingdom, your best avenue for getting started on treks is to work through the British Llama Society at www.britishllamasociety.org.

The group organizes treks for members and is an excellent source of information for outings and llama-related activities in the UK.

Even if you do not have your own pack animal, llama trekking is extremely popular in both the United States and England. Simply use your favorite search engine and type in "llama trekking in the UK" or "llama trekking in the US."

There are numerous companies that will set you up for a day's walking with a llama as a companion or arrange longer hiking outings into often remote and picturesque areas.

Chapter 7 - Llamas as Fiber Animals

Although the alpaca is more traditionally thought of as a fiber-producing animal, llama wool has many of the same characteristics and easily rivals alpaca fiber in quality.

Both alpaca and llama fiber are:

- exquisitely soft
- lightweight
- exceedingly warm
- lanolin-free
- wax and oil free
- hypoallergenic

Llama fiber does not have to go through extensive scouring to be prepared for use and can be used virtually straight off the animal, which makes it highly attractive to fiber artists.

On average a llama's fleece grows 4-6 inches / 10.16-15.24 cm per year and will have a shorn weight of approximately 3-7 lbs. / 1.36-3.18 kg.

The fleece is comprised of an undercoat and guard hairs, with the finest fiber coming from the "barrel" portion of the body. The coarser guard hairs are removed from the soft, downy undercoat during processing.

In shearing the fleece so that the fiber can be used, the barrel of the animal should be blown and brushed free of dirt and debris and then washed and allowed to thoroughly air dry. Remove the fleece in one piece.

Many fiber artists who keep one or two llamas for their own use process the material entirely by themselves. Others seek out llama farmers from whom they can purchase whole shorn fleeces for use in their projects.

The basic steps in processing a llama fleece include:

- removing any remaining debris from the fiber once the fleece has been cut away
- discarding unwanted fiber, like short "second cuts" or remaining guard hairs
- carding the fiber for felting or spinning

Carding is the process of aligning the fibers in the same direction for spinning. Individuals use one of two methods to accomplish this task.

Hand carders look like large slicker brushes used to groom dogs. The fiber is placed between two carders and gently brushed for separating. An alternate method is the drum carder, which has two drums with teeth. A crank transfers the fiber from one drum to the other.

Carded fiber can be used as is for felting by either:

- *wet felting* – In this process, the carded fiber is soaked in hot, soapy water and kneaded to open the scales of the individual fibers and cause them to become interlocked. Once this occurs, the material is rinsed in cold water and worked flat.

- *needle felting* – With a barbed felting needle, the fibers are tangled and compacted in a dry state by repeatedly pushing the needle through the fiber. This is a good method to create small, intricate, decorative pieces.

Carded llama fiber can be spun into thread with either a wheel or a drop spindle. Both provide twist, which creates yarn or thread from the fiber.

The first spinning creates a single ply yarn, which can be combined with two or more other singles to create thicker, stronger thread suitable for weaving, knitting, or crocheting.

As an indication of the quality of llama fiber, the Alpaca Llama Show Association in the United States has competitive classes for:

- *hand spinner's choice* – Participants submit a two ounce / 0.05 kg sample of fleece that is judged on its quality when processed and spun.

- *full fleece* – In this competition, entrants submit the entire fleece from the animal, which is then judged overall quality.

- *walking fleece* – The quality of the barrel fleece is judged while still on the llama, by standards similar to those used in full fleece competition.

Non-ALSA fiber shows of all sorts accept woven, knitted, crocheted, felted, or spun items made from llama wool for judging. Beyond the competitive aspect, however, these items are highly prized clothing and craft items and can be the basis for a thriving cottage industry for individual artisans.

Types of Llama Fiber

The quality of llama fiber varies across four basic types of animals. The finest fiber comes from wooly llamas.

- *Wooly Llamas* have a strong covering of fiber over the entire body with extra density on the neck, head, ears, and down the front legs. Physically they are smaller than other llamas. Their fiber is crimped and fine, will part easily, and has a minimal presence of guard hairs and is thus closest to alpacas in fleece quality.

- *Medium Llamas* have longer fibers growing on the neck and body, but shorter strands on the head, ears, and legs. There are long, rough guard hairs present, so the fleece is referred to as being "double coated."

- *Classic Llamas* have less fiber on the head, neck, and legs with slightly longer hair on the body forming a sort of "saddle." There are some guard hairs on the neck that create a kind of "mane." The fleece is double-coated and lacks uniformity. Overall their bodies are taller and larger.

- *Suri Llamas* are very rare. Their wool coverage is on par with that of the Wooly, but the hair is less fine and forms into rope-like strands. The gene pool to produce this type of animal and fleece is extremely small.

The quality of care a llama receives has a direct effect on the quality of the fiber it produces. Poorly fed animals who are stressed and/or suffering from health problems produce dull fiber with weaker tensile strength.

Marketing Fiber

Very few llama farms can achieve the necessary volume to run a completely self-sustaining llama fiber business. Since all llamas should be sheared for health reasons, however, it is certainly likely that you can sell the harvested fiber to local artisan guilds rather than let it go to waste.

A clean, good-quality fleece commands a price in the range of $3-$7 / £1.8-£4.12 an ounce (28.34 grams). Therefore, a llama fleece weighing 7 lbs. / 3.18 kg would generate around $500-$750 / £298-£442 in income!

Obviously selling good quality harvested fleeces can go a long way toward defraying costs for things like feed, veterinary care, and routine maintenance.

If you are an artisan keeping a few llamas to produce your own materials, you can pool your resources with other artists to make more contacts in the industry and to find venues to market your wares.

There is a certain addictive quality to working with and wearing items made of llama fiber. Most textile artisans say that once they have worked with llama or alpaca, they are spoiled forever to other fibers.

Chapter 8 - Llama Dung for Fertilizer

Llama manure is nitrogen rich and will not burn plants when applied directly. "Llama beans" as the pellets are called are easily processed, highly desirable as an organic fertilizer, and another potentially marketable aspect of llama ownership.

As an added benefit, llama beans do not contribute to the spread of weeds thanks to the camelid's three-chambered digestive system, which completely utilizes all organic materials.

Additionally, neither llamas nor alpacas are known to carry *E. coli o157*, which is responsible for severe illness in humans. Consequently, the dung is safer for use in vegetable gardens than other types of livestock manure.

The excreted pellets are also high in potassium and phosphorous, making the material optimal for plant growth with no added chemicals or negative environmental impact. Good levels of calcium, magnesium, and sulfur are just added extras.

One adult llama produces at least 56 lbs. / 25.4 kg. of beans weekly. It doesn't take long for that to build up! The llamas make collection easy since they use communal dung piles.

Using Llama Beans

There are multiple ways to use alpaca and llama beans as fertilizer including the following:

- Soak a cup of fresh beans (roughly 227 grams) overnight in a gallon of water (3.78 liters). Use the water on houseplants, and keep the beans in your bucket for reuse until they dissolve completely.

- Apply the beans directly to houseplants (spoonful) or at the base of trees (handfuls) in the soil. The fertilizing effect is released when you water the plant and there will be no accompanying odor. (Also the pellets are harmless to pets.)

Llama beans are excellent when mixed with potting soil for transferring plants, and will help to reduce disease and pest problems when applied to lawns.

Storing Llama Beans

Llama beans must be kept moist or they will harden, form a white crust, and break down. To store the material, dampen the beans and keep them well covered.

Check the beans daily and moisten them as needed. Over a period of 30 days, they will break down and take on the appearance of rich peat moss.

Marketing Organic Fertilizer

In offering llama manure for sale, there are a number of points you will want to emphasize to your potential customers who may know nothing about the material.

- The manure has a lower organic matter content than other barnyard manure produced by cows, horses, and sheep.

- Llama dung has high levels of nitrogen and potassium, and average levels of calcium and magnesium.

- Llama manure can be spread directly on plants with no fear of burning them.

- Llamas are not known to transmit *E. coli* to humans.

Animal	Nitrogen %	Phosphorus %	Potassium %
Llama	1.7	0.69	0.66
Chicken	1.0	0.8	0.4
Sheep	0.95	0.35	1.0
Horse	0.7	0.25	0.55
Cow	0.6	0.15	0.45
Pig	0.5	0.35	0.4

As with other aspects of llama ownership, few farms have enough animals to make the marketing of organic fertilizer their sole focus. As a component of the overall operation, however, selling the dung is a complimentary activity that further defrays the cost of owning the animals.

Legal and Licensing Requirements

Clearly if you are using llama dung for your own plants, you are not subject to any legal requirements, but if you want to sell the manure to others, you must make sure that you are following all applicable laws and regulations.

In the United States, contact your local agricultural extension agent or state department of agriculture to discuss specific state and federal laws. These will likely be relative to matters of storage and transportation, and may include provisions to safeguard against the spread of nuisance plants.

Manure sold in its natural state will be subject to fewer regulations than that to which chemicals has been added. Given the naturally rich composition of llama dung, however, such additives are not required for the material to be both desirable and useful.

In the United Kingdom, the handling, storage, and use of manure is determined by EU Control Regulation 1069/2009 and EU Implementing Regulation 142/2011. In general, the use and supply of unprocessed manure is legal so long as there is no danger of spreading a transmissible disease.

For further clarification, contact the Department of Agriculture and Rural Development at www.dardni.gov.uk.

Chapter 9 - Llamas as Livestock Guardians

The use of llamas as livestock guardians is controversial in many circles. Obviously llamas are not going to be able to drive off all predators, and are vulnerable to attack from many animals.

Llamas are, however, capable of defending sheep, goats, and other types of livestock from domestic dogs and coyotes. In practical application, llamas are used most often to guard sheep.

Qualities as Guardians

A single llama will have no difficulty holding off a single predator like a dog but would be badly outnumbered by a pack. It's a complete myth that a llama will use its fighting teeth to attack predators. A guard llama's fighting teeth should be filed down as per normal requirements.

In action, the llama's primary means of defense is to charge, face down, and strike or stomp small predators. Larger intruders, including humans, many be knocked down or even cornered. Llamas can make loud noises, but they rarely give audible warnings of danger.

Llamas have many qualities superior to guard dogs. With good care, they can work for 15-20 years, don't require any special feed, and respect fences. Livestock that might react fearfully to a guard dog will be much more accepting of a llama.

Instinctively, llamas are territorial and highly suspicious of canines. In spite of this fact, however, some llamas will learn to work with guard dogs and to tolerate pet dogs that do not pose a threat to the animals being guarded.

The Guard Bond

The longer a llama stays with a herd or any kind of livestock, they develop a feeling of responsibility for the animals and will even stand guard when females are giving birth.

In some instances, the bond becomes so strong that a llama will react aggressively to any people who come near their charges. This can be both inconvenient and potentially dangerous, so it's best to confine the llama before performing any routine chores with the other livestock.

The guard bond that develops can be with the specific animals or with the general area in which the animals are being kept. In some instances the llama's guard bond does not develop at all until the livestock in its care begin to lamb, kid, or calve.

Selecting a Guard Animal

Conventional wisdom once held that single llamas be used for guarding duties, but now the trend is toward using pairs of females.

By nature the females are collaborative and cooperative. Their nature disposes them to believe that there is safety in numbers, and they seem to have an almost maternal attitude toward the livestock in their care.

There is no requirement that a guard llama be white. Some sheep ranchers simply prefer to use white animals so hunters do not mistake them for game.

In general:

- Males are the most territorial by nature, but can be prone to fighting with other males rather than attending to their duties. They may also attempt to breed with intact ewes and cause serious injury.

- Females are more likely to be good guardians due to their maternal instincts.

- Geldings retain their territorial instinct, but may also be subject to breeding triggers. They may also injure intact ewes and are thus not suitable as guard animals.

The average gelding will not attempt to breed with other species in season, but it is still possible for all intact males and geldings to be indiscriminate about their potential sexual partners irrespective of species.

Even a gelding that has been a guardian for several years can suddenly exhibit such behavior. When all the risks are weighed, the best option is to use pairs of females. In summary, desirable qualities for a guard llama include:

- age of 18 months at minimum
- females that are not pregnant
- confident demeanor
- respectful of humans
- receptive to halters and leads
- allows examination of body and feet
- allows grooming and shearing
- alert and aggressive toward strange dogs
- good formation with no physical faults

Chapter 10 – Showing Llamas

In the United States, llamas are popular show animals for enthusiasts of all ages, including participants in high school 4-H and Future Farmers of America (FFA) programs. Under the auspices of these groups the animals may be shown in conjunction with county fairs and local and regional livestock shows among other venues.

Events that are sanctioned by the International Lama Registry and the Alpaca Llama Show Association in the United States and by The British Llama Society in the UK attract participation by top-notch breeders affording attendees the opportunity to see some of the finest animals in the industry.

The best way to learn how to show your animals is to both attend llama shows and to sign up for showing clinics. There are various classes with different requirements, but in general judges are looking for the following points:

- **Head and Neck** – An alert expression, uniform shape to the ears, wide eyes that are well set and bright. Proper alignment of the jaw and normal teeth.

- **Body** – A straight topline that continues off the end of the tail. (Tails that are too low indicate potential crossbreeding with alpacas.) Proper muscling and condition, proper distance between the front and back legs, four teats on females and two visible and uniform testicles on males.

- **Legs** – Good bone density with straight alignment, toes facing forward and correct alignment of the pasterns. Clean feet and properly trimmed toenails.

- **Fleece** – Uniform density, fineness of the fiber, overall luster and health. Quality of the grooming.

- **Balance and Movement** – Fluid gait with graceful movement and no short, choppy steps. Well-proportioned and symmetrical balance throughout. Proper conformation.

- **Attitude or Disposition** – Good attitude with an obvious desire to please, manageable disposition.

Clearly there are additional requirements in events like the Halter Class, Showmanship Class, or Obstacle Class. For instance:

- Animals in the Halter Class must be taken around the ring at a brisk walk with the llama being led and positioned in precise ways.
- In Showmanship Classes, the handlers are the ones being evaluated, not the animals.
- Obstacle Classes include water, stairs, ramps, and bridges the animals must negotiate and are flexible and often fun affairs.

Some shows also include fiber judging where whole fleeces or small samples are submitted for competitive evaluation.

If you are interested in mastering the art of showing your animals, the experience can garner impressive credibility for a breeding program, and is a superb activity for young people who especially enjoy the training and activity involved in obstacle work.

The finer points of learning to show llamas is out of the intended purview of this book, but you can find more information on shows to enter and preparatory classes in which to enroll by contacting your state, regional, or international llama association.

Chapter 11 - Frequently Asked Questions

While I recommend reading the entire text of this book and working hands on with llamas under the mentorship of someone knowledgeable in their care, the following are some of the questions I am asked most frequently about these beautiful animals.

Can a llama be a good pet?

Llamas absolutely make good pets. You should think of them as "field pets," however, not the kind of animal that lives with you and plays in the backyard. Llamas are gentle, fun, and fascinating, but they're not going to curl up on the couch with you!

Llamas can, however, come into a house for brief periods because they tend to use communal dung heaps and specific "latrine" areas by choice.

They are also excellent companions for walking, packing, hiking, and other outdoor activities in areas where they are allowed. See the chapter on packing with llamas for more information.

What do llamas eat?

For the most part llamas are perfectly happy to graze on grass in the field and to eat supplemental hay as needed. They can have the occasional carrot or apple peel as a treat, but their digestion can be easily subject to upset.

Be sure to read the chapter on llama husbandry carefully to really understand how to feed your animals properly and to ensure they receive proper nutrition.

Are llamas good with children?

In general llamas get along well with people of all ages, but interestingly, they do seem to have a special affinity for the very young and the very old. For this reason, they are often used as therapy animals, even going on hospital visits.

How do llamas do with other animals?

Llamas will live quite peacefully with other livestock and actually seem to enjoy being companions for horses, donkeys, and ponies.

Female llamas do well with sheep and goats, but care should be taken with intact males and geldings. (See the section on guard animals for more detailed information.)

Although llamas are naturally suspicious of all canines, they can be quite accepting of dogs so long as there's no aggression on Fido's part.

Are llamas good guard animals for other livestock?

Some llamas, if carefully selected for gender and temperament, will protect sheep, goats, and even poultry from predators. Llamas are not impervious to attack from aggressive predators, however, and it is a mistake to assume that all you have to do is put a llama out with your livestock. See the section on llamas as guard animals to understand more about finding a suitable animal for this work.

Can you ride a llama?

A very young child might be able to ride a well-trained adult llama in a pen, but overall llamas are not large or robust enough to be ridden. They can be used to pull small carts however, and are often used as pack animals on trekking expeditions.

Would you say llamas are friendly?

I think it would be more accurate to say that they are curious by nature and that can translate to being friendly over time as they become accustomed to specific people. By nature, llamas are shy and somewhat independent. They are, however, intelligent and highly trainable.

Do llamas make a lot of noise?

Llamas are very quiet. They occasionally hum, and they do have a braying alarm call often described as an "oynk," which they will use when they detect an intruder. Males make loud gurgling sounds when they are courting females. Otherwise, you'll never even know they're around!

Are llamas really easy to train?

Yes, llamas are quite intelligent. They respond well to praise and rewards and often require very few repetitions of a task to "get" what you want them to do. When you are teaching a llama a routine with several steps, like accepting a pack, you can often take up the lesson one day exactly where you left off the previous day with only minor repetition.

Are llamas difficult to breed?

It is rare for llamas to require assistance either in mating or in giving birth. Their gestation period is 11.5 months. Females usually give birth to a single cria annually after they reach 18 -24 months of age. Males, however, are not fertile until 24 - 30 months of age.

What is a llama's lifespan?

Many well-cared-for llamas live into their twenties or even early thirties. A good average lifespan number is around 15 years of age.

Are llamas healthy animals?

Llamas are indigenous to the high mountain plains of the South American Andes. They have been domesticated there for thousands of years in a climate where survival of the fittest is the rule, not an evolutionary theory. The result is a species that is disease resistant and very hardy.

Is keeping llamas expensive?

Llamas are not expensive. They eat less than sheep. They rarely need to see a vet, and they are very easy on the pastures where they graze. Add in minimal shelter needs and non-challenging behavior with fences, and they are likely the best behaved of all types of larger livestock.

Also, as you will come to understand from the text, there are numerous ways to offset ownership expenses from selling the animal's fleece to even marketing its dung!

How much land do llamas need?

Estimates vary, but an acre of land with good grass cover will support at least 3 llamas, potentially four. You should, however, have enough land to rotate pastures.

Llamas don't pull grass up by the roots, preferring to just nip off the tender top shoots, but the growth will still need recovery periods.

What type of shelter and fences will my llamas need?

Simple three-sided livestock shelters and fairly standard fences can be used with llamas. Please see the chapter on husbandry for a detailed discussion of each of these topics.

How often do llamas need to be sheared?

Remember that llamas are indigenous to a high mountainous area. They do well in temperate climates, but can easily be subject to heat stress. (See the health chapter for a full discussion of this topic.)

Even if you are not planning to harvest their wool for use as fiber, shear your llamas in the spring before the weather gets really hot. Always make sure your animals have access to shade and water.

How old should llamas be at purchase?

Sale practices vary by breeder, but in most cases young llamas are not available for sale until they are 12-15 months of age. If the youngsters are going into a new herd, however, they can be sold at 8-10 months.

Are llamas raised as meat animals?

Llamas are not farmed for their meat in the United States or Europe, but in Bolivia, they are used as food animals.

What colors are common in llamas?

Llamas are seen in all color gradations between black and white. Some are solid colors, while others are patchy or spotted.

Is a license required to keep llamas?

Llamas are considered to be domestic livestock in both the United States and the UK, so no license is required.

Do llamas spit?

It is very rare for llamas to spit on humans. You'd have to have one seriously ticked off animal for that to happen.

They may, however, spit at one another to settle issues of territoriality and to enforce the pecking order in the herd.

Can I keep just one llama?

If you really want your llamas to be happy and to thrive, plan on keeping a pair at least. These are herd animals, and they don't enjoy living alone, often reacting with aggressive and unruly behavior.

Is it hard to find a vet to treat llamas?

Any large animal vet is capable of treating a llama since they don't have any unusual disease. They really are not all that different from other types of livestock.

More and more vets have experience with llamas or can easily find a colleague with whom to consult. It's a good idea to find out about veterinary services in your area before acquiring your animals, but this is hardly an impediment to ownership.

Chapter 12 – Breeder Directory

United States

California

Fallen Oak Llamas
www.fallenoakllamas.com

Wild Oak Llamas
wildoakllamas.com

Colorado

Rockwood Llamas
www.rockwoodllamas.com

Georgia

Rock Creek Farm & Llama Co.
www.rockcreekllamas.com

Kentucky

S & S Camelid Co.
www.snscamelidco.com

Michigan

Star Llama Company
www.starllama.com

Mississippi

PKS Llamas
www.pksllamas.com

Missouri

CriVen Llamas
crivenllamafarm.com

Montana

M and M Llama Ranch
www.mandmllamas.com

New Jersey

Spruce Lane Llamas
www.sprucelane.com

Vermont

Autumn Mountain Llamas
autumnmountainfarm.com

United Kingdom

Berkshire

Ordellamas
www.ordellamas.co.uk

Cheshire

Hill View Llamas
www.hillviewllamas.co.uk

Cornwall

Calamankey Llamas and Guanacos
www.calamankey.com

Llama Lland Llamas
www.llamalland.com

Devon

Ashwood Llamas
www.ashwoodllamas.co.uk

Watertown Llamas
www.watertownllamas.co.uk

Hampshire

Dunley Park Llamas
www.dunleyparkllamas.com

Herefordshire

Golden Valley Llamas
www.oldkingstreetfarm.co.uk

Northants

Catanger Llamas
http://www.llamatrekking.co.uk/

Sussex

Bluecaps Llamas
www.bluecapsllamas.co.uk

Wales

Bremia Farms
www.llamas-bremia.co.uk

Afterword

Back in my Colorado early "mid-life crisis" days, my neighbor's wife was right to call me Professor Old McDonald in a waggish reference to the farmer of the same name.

I really didn't have any business living by myself in a tiny little cabin, a fact brought home to me by my first winter in the mountains. My neighbors not only gave me a job when I really did need the money, but they probably kept me alive.

The llamas I mistakenly took to be "long necked sheep" were a big part of those years for me. I was about as qualified to care for livestock as I was to live in a mountain cabin, but even the llamas seemed forgiving about my youthful optimism, no doubt inspired by one too many readings of *Walden*.

The extent of my civil disobedience up there in my solitary cabin was dissatisfaction with the politics of academia. My planned "great American novel" was intended to feature a thinly veiled autobiographical protagonist who put one over on the stiffs in the ivory tower.

That didn't happen, because between the chopping wood, porch sitting, fly-fishing, and llama pronking, a strange peace came over me.

There are many reasons to have animals like llamas in your life, many of them tied to a potential for profit, but as I was researching this book, I was struck by the number of times I ran across references to how peaceful it is just to walk with llamas. I have no doubt that's why llama trekking has become popular. I have walked with llamas and they are excellent and undemanding company.

During the two years I spent in the mountains, I learned enough about llamas to truly appreciate their uniqueness, and I learned enough about myself to finally pack up my tiny cabin and come back down to the world. I actually credit the llamas with part of that awakening. They are incredibly content to be in the world just as they are.

If I ever go up into the mountains again, or have a little house out in the country as I always say I'm planning to do, I would love to have a pair of llamas grazing in my pasture.

I wouldn't need them to guard my sheep since I don't have any intention of owning them, and I can't see myself becoming a llama dung farmer or hiking up into the wilderness.

For the most part, I'm an armchair adventurer. But I can see myself walking the field as the sun sets talking about my day in the company of a convivial llama.

When a llama gets used to you and trusts you, I swear they listen attentively and seem poised to offer you their particular wisdom on the ways of the world. I'm often asked if they make good "pets," and I typically suggest the word "companion" as an alternative.

I've tried in this book to give you a sense of how to enter the world of llamas by finding a mentor and understanding enough about these South American camelids to start asking the right questions. There's absolutely no replacement for hands-on time when you're a llama newbie.

As I said in the introduction, there's a reason this book isn't called something like "The Encyclopedia of Llama Care." I didn't set out to write the definitive guide, but I do hope I've either sated or piqued your curiosity about what life with a llama would be like.

Accept the fact now that you will make mistakes, but remember the infinitely correctable nature of ignorance. You will learn, and for the most part the llamas will be patient with you while you're doing it.

My memories of my sojourn in Colorado are made complete by those llamas. They were intelligent and curious, very gentle, and endlessly interesting.

These are not animals you should go out and adopt on a whim. They aren't demanding, but they do require a set standard of care you should be confident of meeting well in advance of coming home with your first pair of llamas.

If I've taught you enough about llamas that you are now eager to bring them into your life, I congratulate you on the adventure that lies ahead.

Relevant Websites

Please note that the following websites were extant in mid-2014 at the time of this writing. Due to the ever-changing nature of the Internet, no guarantee can be made that these addresses will be valid in the future.

Alpagas et Llamas de France & Registry (l'AFLA)
www.alpagas-Llamas-france.org

Alpaca Owners Association, Inc. (AOA)
alpacainfo.com

Alpaca Llama Show Association
www.alsashow.org

American Miniature Llama Association
www.miniaturellamas.com

British Llama Society
www.britishLlamasociety.org

Camelid Identification System (CIS)
www.cisdna.org

Canadian Llama and Alpaca Association and Registry (CLAA)
www.claacanada.com

International Camelid Institute
www.icinfo.org

International Lama Registry
secure.lamaregistry.com

Llama and Alpaka Register (Austria)
www.Llamas.at

Llama and Alpaca Registries Europe (LAREU) (Switzerland)
www.lareu.org

Llama Association of North America
www.lanainfo.org

Llama Futurities Association
www.thelfa.org

Llama Nation
www.llamanation.com

Glossary

Alarm Call - An alarm call is a sound typically made by male llamas that signifies their perception that the herd is being threatened. The sound has been compared to an engine turning over or to a turkey call.

Artificial Insemination (AI) – Artificial insemination is the procedure whereby a human inserts semen from a male llama manually into the uterus of a female. Although widely used in many species of livestock, artificial insemination is rarely successful with llamas.

Banana Ears – The term "banana ears" refers to the set of a llama's ears that curve up and inward approximating the size and shape of a banana.

Berserk Male Syndrome – The condition Berserk Male Syndrome occurs in male llamas that have improperly imprinted on humans. At puberty, the llama becomes physically aggressive toward people. This behavior can rarely be corrected.

Body Score – A body score is a numerical value to rate a llama's weight. The scale runs from 1 to 9, with 5 being optimal. A rating of 1 would indicate an emaciated animal, while a 9 would indicate obesity.

Bone – The term "bone" describes the size of a llama's skeletal frame. Animals with a large frame are said to have "a lot of bone."

Colostrum – Colostrum is a female llama's first milk. It is rich in antibodies essential to the health of a newly born cria's immune system.

Cria – The term "cria" is the proper reference for a young llama from birth to the point when the animal is weaned from its mother's milk.

Dam – The appropriate term for a llama's female parent.

Dung Pile – Dung piles are areas where llamas choose to urinate and defecate. Typically there will be several dung piles in each pasture.

Dust Pile – Dust piles are bare areas of ground where llamas enjoy rolling.

Embryo Transfer (ET) – Embryo Transfer involves the removal of an early embryo from one female llama to another. The process is rarely successful in llamas.

Forage – Forage is that part of a llama's diet that is high in fiber but less dense in energy and is comprised of grasses, hay, and legumes.

Gait – A gait is a set type of locomotion or movement. Llamas exhibit five types of gait: walk, pace, trot, gallop, and pronk.

Gallop – A gallop is a gait in three beats. All four feet never strike the ground together. It is the fastest of the camelid gaits.

Get of Sire – At a llama show, the "Get of Sire" is a class where three animals with the same sire and a minimum of two different dams are exhibited as a group. The judge looks for consistency and influence derived from the sire.

Going Down – The phrase "going down" refers to the behavior of a female llama when she drops into the kush position signaling her receptivity to mate with a male.

Herdsire – The herdsire is a male llama used to breed females. He may also be referred to as a "stud." Often herdsire's services are sold for the collection of "stud fees" to llama owners who do not have intact males on their farms.

Humming – Humming is the sound llamas make when they are stressed, tired, curious, concerned, or even just hot. Mother llamas hum to their offspring at birth to encourage and guide them to their first feeding, and even at so young an age, the cria will answer with a hum of its own.

Knock-kneed – A llama is said to be knock-kneed when the front knees are angled inward toward one another. The medical term for the condition is *carpal valgus*. This fault prevents correct movement and is the precursor of degenerative joint disease.

Kush – The sternal reclining position adopted by llamas is called the "kush." The word is also used as a command to encourage a llama to move into this resting position for transport or other purposes.

Lama – The term "lama" is used for both llamas and alpacas as both belong to the genus *Lama*. It is not a misspelling of the word "llama" as is commonly believed by people new to the camelid world.

Maiden Female – Young female llamas that have not yet been bred to a herdsire are referred to as "maiden females."

Open Female – An open female llama is one that is not pregnant and is thus available for breeding provided she is of the correct age.

Overconditioned – The term "overconditioned" is used in the show ring to indicate in a polite way that a llama is overweight.

Pace – A pace is a gail with two beats in which the front and rear legs on the same side move forward or back at the same time. This achieves a medium, but relatively unstable motion.

Packer – A llama capable of taking large loads over long distances is said to be a "packer." Typically these are animals with a light coverage of wool and larger bodily frame and conformation.

Paddling – "Padding" refers to faulty motion on the part of a llama. The front feet swing away from the body as the leg comes forward. The motion is reminiscent of that of a bulldog and is a consequence of an overly wide chest. The fault may be genetic in nature, or the result of the animal being overweight.

Potty Pile – "Potty pile" is a common term used in place of dung pile to indicate a communal location where a herd of llamas urinate and defecate. Typically there are several such piles in a pasture.

Produce of Dam – In llama competition, the "Produce of Dam" class involves the exhibition of two llamas with the same dam but different herdsires. The animals are shown as a pair so the judge may evaluate the dam's consistency in influencing the conformation of her offspring.

Preemie – The use of the term "preemie" is identical in humans and llamas and in the animals refers to a cria born prematurely.

Pronking – Both adult and juvenile llamas "pronk" when they play. The action is an upward bounce with legs held stiff. It is also a gait used to evade predators.

Rolling – Llamas regularly lay on their side and roll over half-way several times. The motion keeps their fiber open to allow for the formation of insulating air pockets.

Sickle-hocked – A llama that is sickle-hocked is one in which the hind feet are placed too far forward thus creating the appearance of sickle-shaped hindquarters when the animal is viewed from the side.

Sire – The sire is the male parent of a llama.

Stud – A stud, also known as a herdsire, is an intact male llama used to breed females.

Three-in-One – The term "three in one" refers to the sale of a pregnant female llama with her unweaned cria. The buyer pays one price for the female, the cria, and the unborn baby.

Tipped Ears – Llamas that do not have completely erect ears are said to have "tipped ears." The cause of the malformation is weak cartilage at the ear tip caused by a genetic abnormality, premature birth, or frostbite.

Topline – When viewed from the side a "topline" is the line of the llama's back, which should be level from the withers to the tail.

Trot – A trot is a gait with two beats in which the opposing front and rear limbs move forward simultaneously. It is a stable gait with medium speed.

Underconditioned – Llamas that are underweight are said to be "underconditioned."

Walk – A walk is a gait with four beats in which three feet remain in contact with the ground at one time. It is the slowest of all gaits exhibited by llamas.

Weanling – A llama under one year of age that has been weaned from its mother.

Winging – Winging is a term for faulty movement in llamas in which the front foot swings away from the midline when the leg is brought forward. It is typically associated with the condition of being knock-kneed.

Woolies – Llamas that have a very heavy covering of fiber or wool are sometimes referred to as "woolies."

Yearling – As the word implies, a yearling llama is one that is one year of age and not yet two.

Index

Lightning Source UK Ltd.
Milton Keynes UK
UKOW06f1828240915

259219UK00005B/320/P